Lecture Notes in Control and Information Sciences

Edited by M. Thoma

Vol. 43: Stochastic Differential Systems
Proceedings of the 2nd Bad Honnef Conference
of the SFB 72 of the DFG at the University of Bonn
June 28 – July 2, 1982
Edited by M. Kohlmann and N. Christopeit
XII, 377 pages. 1982.

Vol. 44: Analysis and Optimization of Systems
Proceedings of the Fifth International
Conference on Analysis and Optimization of Systems
Versailles. December 14–17, 1982
Edited by A. Bensoussan and J. L. Lions
XV, 987 pages, 1982

Vol. 45: M. Arató
Linear Stochastic Systems
with Constant Coefficients
A Statistical Approach
IX, 309 pages. 1982

Vol. 46: Time-Scale Modeling of Dynamic Networks
with Applications to Power Systems
Edited by J. H. Chow
X, 218 pages. 1982

Vol. 47: P. A. Ioannou, P. V. Kokotovic
Adaptive Systems with Reduced Models
V, 162 pages. 1983

Vol. 48: Yaakov Yavin
Feedback Strategies for Partially
Observable Stochastic Systems
VI, 233 pages, 1983

Vol. 49: Theory and Application of Random Fields
Proceedings of the IFIP-WG 7/1
Working Conference
held under the joint auspices of the
Indian Statistical Institute
Bangalore, India, January 1982
Edited by G. Kallianpur
VI. 290 pages. 1983

Vol. 50: M. Papageorgiou
Applications of Automatic Control Concepts
to Traffic Flow Modeling and Control
IX, 186 pages. 1983

Vol. 51: Z.. Nahorski, H.F. Ravn, R.V.V. Vidal
Optimization of Discrete Time Systems
The Upper Boundary Approach
V, 137 pages 1983

Vol. 52: A. L. Dontchev
Perturbations, Approximations and Sensitivity Analysis
of Optimal Control Systems
IV, 158 pages. 1983

Vol. 53: Liu Chen Hui
General Decoupling Theory of Multivariable
Process Control Systems
XI, 474 pages. 1983

Vol. 54: Control Theory for Distributed
Parameter Systems and Applications
Edited by F. Kappel, K. Kunisch,
W. Schappacher
VII, 245 pages. 1983.

Vol. 55: Ganti Prasada Rao
Piecewise Constant Orthogonal Functions
and Their Application to Systems and Control
VII, 254 pages. 1983.

Vol. 56: Dines Chandra Saha, Ganti Prasada Rao
Identification of Continuous
Dynamical Systems
The Poisson Moment Functional
(PMF) Approach
IX, 158 pages. 1983.

Vol. 57: T. Söderström, P. G. Stoica
Instrumental Variable Methods
for System Identification
VII, 243 pages. 1983.

Vol. 58: Mathematical Theory of
Networks and Systems
Proceedings of the MTNS-83 International
Symposium
Beer Sheva, Israel, June 20–24, 1983
Edited by P. A. Fuhrmann
X, 906 pages. 1984

Vol. 59: System Modelling and Optimization
Proceedings of the 11th IFIP Conference
Copenhagen, Denmark, July 25-29, 1983
Edited by P. Thoft-Christensen
IX, 892 pages. 1984

Vol. 60: Modelling and Performance
Evaluation Methodology
Proceedings of the International Seminar
Paris, France, January 24–26, 1983
Edited by F. Bacelli and G. Fayolle
VII, 655 pages. 1984

Vol. 61: Filtering and Control of Random
Processes
Proceedings of the E.N.S.T.-C.N.E.T. Colloquiur
Paris, France, February 23–24, 1983
Edited by H. Korezlioglu, G. Mazziotto, and
J. Szpirglas
V, 325 pages. 1984

For information about Vols. 1–42 please contact your bookseller or Springer-Verlag.

Lecture Notes in Control and Information Sciences

Edited by M. Thoma and A. Wyner

112

V. Kecman

State-Space Models of Lumped and Distributed Systems

Springer-Verlag Berlin Heidelberg GmbH

Series Editors

M. Thoma · A. Wyner

Advisory Board

L. D. Davisson · A. G. J. MacFarlane · H. Kwakernaak
J. L. Massey · Ya Z. Tsypkin · A. J. Viterbi

Author

Vojislav Kecman
Faculty of Mechanical Engineering
and Naval Architecture
University of Zagreb
Dj. Salaja 5
Zagreb 41 000
Yugoslavia

ISBN 978-3-540-50082-7   ISBN 978-3-540-45922-4 (eBook)
DOI 10.1007/978-3-540-45922-4

Offsetprinting: Mercedes-Druck, Berlin

2161/3020-543210

PREFACE

This work is devoted to the science and art of the mathematical description of dynamics found in various technical processes and systems with spatially lumped and distributed parameters. The reader will find basic and indispensable knowledge here, and master the techniques and secrets for devising a dynamic mathematical model making use of the fundamental laws of nature: the laws of conservation of mass, energy and momentum. But this is the book's lesser part; it is not merely designed to provide insight into the general and special aspects of the mathematical modeling of thermal, mechanical, hydraulic, pneumatic and chemical dynamic systems. Other goals and purposes also motivated the volume you are holding.

Firstly, to show the real relationships between the process variables, physical and geometrical properties and the structure, coefficients and parameters of mathematical models, i.e. to show the connections between the physical world and abstract concepts of eigenvalues, poles and zeros of state-space models and their Laplace representations in the s-domain.

Secondly, to present how indispensable assumptions in the modeling process are mapped onto a mathematical structure and the parameters of such dynamic (and static) models.

Thirdly, to show connections, similarities and differences between dynamic models of lumped and distributed systems, with particular attention to the question of discretization (i.e. model reduction) and its consequences in the world of both model structure and parameters, and in the representation of real dynamic properties. The limitations of analytical tools are pointed out in even the most elementary and simplest distributed systems.

To realize these tasks it is possible (even necessary and useful) to analyze the dynamics of physically completely different processes using the same unified procedure and method. From this stem two important characteristics of the book as a whole - a consistent approach to the dynamics of different systems using the concepts of dynamic variables (accumulated variable, effort, flow and general acceleration) and dynamic coefficients (of capacity, resistance and inertia), and state-space representation of all the obtained lumped models. The latter feature could be of particular use for control system analysts and designers.

The book is divided into three chapters (followed by an Appendix with a presentation of the basic mathematical tools used in the book).

The introductory chapter outlines the overall concept of the whole book. It defines the basic ideas, gives a process classification, defines the focal dynamic problem and introduces the basic concepts of dynamic variables and coefficients in different technical systems. The rest of the book is then divided according to dependence of system parameters upon the spatial coordinate.

Chapter 2 considers processes with lumped parameters. The process of mass storage, fluid flow, heat transfer and mechanical processes of rigid body motion are then thoroughly examined. It is pointed to their basic and common dynamic properties (named here as proportional, integral and derivative). The second part of this chapter leads to distributed systems and is devoted to the modeling of complex systems of a higher order. The last example of this chapter deals with the problem of the discretization of partial differential equations (PDE) and presents an introduction to Chapter 3 which considers the question of dynamic models of distributed systems.

Following the common practice in mathematical physics, this third and last chapter is divided into three sections. The first one deals with mass and energy transportation systems which are described by hyperbolic PDE of the 1st order. The second section is devoted to systems with equalization, which are described by parabolic PDE, while the last one deals with systems with periodic state changes (described by hyperbolic PDE of the 2nd order). This chapter provides a completely analytical (and just partly numerical) study of all the important aspects of the dynamics of distributed systems with particular emphasis on connections and comparisons with lumped systems. The last two sections have a monographical character, but are written in quite a readable way.

Concerning the notation, it must be said that it was at times very difficult (impossible, in fact) to avoid using the same notation, symbols and subscripts for physically different variables. This results from the fact that this book contains many variables and coefficients from different technical disciplines which developed independently of each other and have their own separate terminology and symbols. So it was, for the present, impossible to be completely consistent in this regard. Finally, it must be said that in SerboCroatian this book is called "Process Dynamics" and some of the comments in the introduction refer to this title connected with a course of the same name held at the Faculty of Mechanical Engineering and Naval Architecture, University of Zagreb, part of the undergraduate and postgraduate curriculum.

The book was written for scientists, practising engineers and students interested in the analysis of system dynamics. Experience has shown that the volume is of special value to analysts and designers of control systems in many disciplines of engineering. The first two chapters can be of great use as a textbook for subjects from the field of dynamics and control systems in university undergraduate courses, while the third chapter is intended for more detailed graduate study. Having this in mind, every section of the book ends in many solved numerical examples. This can be of great use in the continuing education and home-study of all those who are concerned with this fast-developing field.

Finally, it must be said that no book gets its final form through the efforts of its author alone. There are always many others deserving credit for stimulation, inspiration, support and help and without whose efforts books would not reach their readers. In the case of this book I wish to extend my most heartfelt gratitude to Professor Tugomir Šurina, University of Zagreb, for his unending support during research in this field, and to Professor Petar V. Kokotović, University of Illinois, for support in the promotion of the English edition of this book. The extensive and successful work on the English translation was performed by Ms. Nikolina Jovanović, whom I also thank. To Ms. Ellen Elias-Bursać, M. A., thanks for her work on the language editing of the English text. The merits for well-drawn figures and patient typing belong to Zvonko Grgek and Dragica Špoljar, respectively, as do my thanks. Finally, after everybody else had finished their work, the exacting job of setting the whole text on a personal computer and editing was performed by Đorđe Tasevski, M.Sc., for which the writer of these lines is sincerely thankful.

Naturally, the list of institutions, colleagues and friends who contributed to the writing of this book is much longer than the always limited space allotted to such prefaces allows. I will warmly and sincerely thank all of them in person. Thus this foreword ends.

Let the book begin.

Zagreb, June 1988.   The Author

CONTENTS

VIII

INTRODUCTION

1.1 BASIC CONCEPTS, NOMENCLATURE AND CLASSIFICATION

Until recently, dynamics (from the Greek word dynamis - force, power) was a clearly defined technical discipline and science, and the meaning of the word did not demand special explanation. It was the name of a science which formed a part of the wider concept of mechanics and studied the spatial motion of a particle or a rigid body due to force. The foundations of dynamics were Newton's laws, which have since the years of their establishment also been applied outside the field of classical mechanics, and in time used more and more in the study of motion of continua (elastically and plastically deformable bodies, liquids and gases), which contributed to the development of fluid dynamics, gas dynamics or aerodynamics, and the theory of elasticity. Differential equations, ordinary and partial, were and still are the characteristic mathematical tool used to describe the process of motion. In time, the field of disciplines, in which differential equations appear in mathematical representations, expanded. This mostly occured in problems of control in various technical and nontechnical processes - mechanical, electrical, heat, flow, chemical, physical, biological, sociological and the like. As the field of described and analyzed processes grew, already existing terms were kept, but their meaning changed and became more generalized. Today coordinates are not only the geometrical, spatial coordinates describing the position of a particle, but all the indicators of the process under observation (temperature, pressure, concentration, speed). Moreover, when we say motion, we no longer mean only displacement or a change of the rigid body's spatial coordinates, but any time change in coordinates defining the state of the analyzed process.

In short, a more extended definition may be given by which dynamics is the science which deals with time changes (variations) of state, or which studies the changes of state with time. Its main feature is that time t is one of, or the only, independent variable in the mathematical model that describes such changes. If time t is the only independent variable the dynamics will be described by ordinary differential equations (ODE). If the independent variables include spatial or some other coordinates (for example, age in sociological processes) besides time, the dynamics will be described by partial differential equations (PDE).

The subject of this book is mathematical modeling and analysis of process dynamics. Here we consider a process (from the Latin processus - progressing, growing) to be an arbitrary qualitative and quantitative change in time, i.e. a time change, reshaping or transformation in quality and quantity. The word dynamics in the term implies that our interest lies in the time profile of these changes, their mathematical description and the analysis of the obtained model.

The word change must be considered in its most generalized meaning, it can mean changes in the shaft diameter during lathe manufacturing, temperature and pressure changes at the exit of a steam generator, density changes in a chemical reactor, an airplane's changes in altitude, or changes in the shape of plants, changes in human relations and so on. This short list also shows something else. Each of these processes takes place in some space, element, object, process unit or plant. Thus it is not unusual that there are many books with different names that treat, for instance, the dynamics of objects, the dynamics of systems, or even more specifically, the dynamics of chemical reactor or energy plants. Their names include small terminological, but also conceptual, differences and traps, and we would not get far in an effort to solve them. Here, and for now, this is not our purpose. In connection with the name of this book, the following must be said. It was not reached by chance. Moreover, it was purposely chosen to study the dynamics of specific and different processes with a unified approach, method and conceptual tools. In technical devices in which only one process takes place it is correct to equalized the term of object dynamics with the term of process dynamics. One of the simplest examples where this is fulfilled is a liquid storage tank, in which only the process of storing (accumulating) a mass of fluid takes place. An indicator of the state in that object is head H, and the use of the term dynamics of the liquid storage tank or dynamics of the process of liquid storage will not cause serious misunderstanding.

This is not the case in more complex technical objects in which several different processes take place simultaneously with usually unlike dynamic characteristics. Thus it would, for example, be more correct to use the term dynamics of (heat and/or of flow) processes in a steam generator instead of only the dynamics of a steam generator, which is, incidentally, more usual. In this latter case the impression may be gained that only one process takes place in the steam generator whose dynamics we wish to examine, while there is in fact a whole spectrum of dynamically different phenomena, from the slower processes of heat transfer on superheater surfaces to the rapid, high-frequency phenomena of hydraulic shocks in evaporator pipes. (The concepts of slowness and speed are of relative character, and what we here call fast processes of hydraulic shocks would certainly not be fast in comparison with many mechanical oscillatory processes and especially with processes in electrical circuits.)

Table 1.1-1

		PROCESS CLASSIFICATION		
		according to	into	
A	mutual dependence of variables	linear		nonlinear
B	parameter change with time	time invariant		time variant
C	dependence on spatial coordinate	lumped		distributed
D	processing of material or energy	flow	charge	discrete
E	randomness of variables	deterministic		stochastic
F	state variable change with time	static		dynamic

The above definition of a process is obviously somewhat wider than this book could be expected to cover, and it would be of use to narrow it down to the field that will be the subject of study here. Furthermore, a classification of processes according to some of their significant characteristics can be a good introduction into fields that are to be investigated in the coming chapters. Besides being electrical, mechanical, heat, hydraulic, pneumatic, chemical and so on, each of these processes can be subjected to a classification like the one in Table 1.1-1.

What each of these terms includes will be briefly described in the following lines:

A

Linear processes are all processes in which the relationships between process variables of input u, state x and output y can be described by linear mathematical expression of the following type

$$y = a_1u + a_2, \quad x' = a_1x + a_2u + a_3, \quad x' = a_1u_1 + a_2u_2 + u_3 + u_4 \qquad (1.1-1)$$

(On the concepts of input, state and output see in the Appendix.)

Coefficients a_i are constants, and variables x, x', y and u are time variant functions and could, consequently, be written x(t), x'(t), y(t) and u(t). If the dependence between process variables are as follows

$$y = a_1ux, \quad x' = \frac{x}{u}, \quad x = \sqrt{xu}, \quad x' = u \sin x, \qquad (1.1-2)$$

or can be represented by some other nonlinear equations, such processes will be called nonlinear.

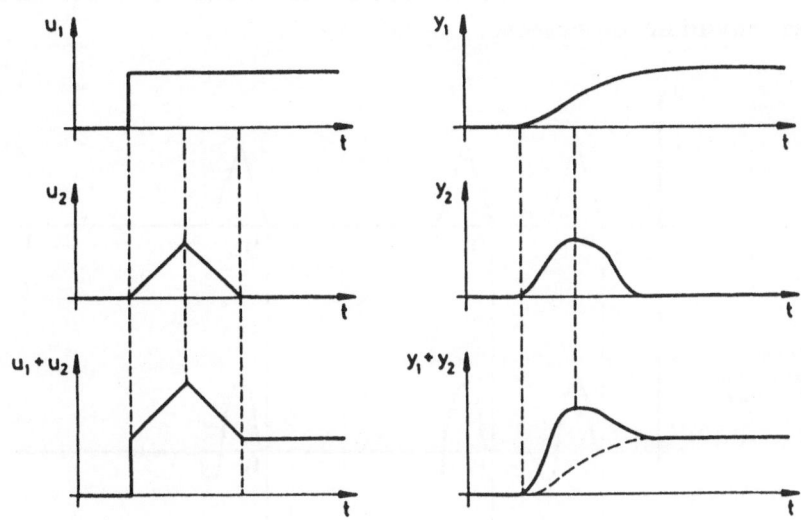

Fig. 1.1-1 The principle of superposition

In practice most processes are of a nonlinear character, and in order to facilitate analysis their description is very often translated into linear form. This will be one of the noticeable characteristics of this book. One of the key linear properties (the one which has made linear models the most desirable) is that the principle of superposition is applicable. This principle can be expressed in words as follows:

> if input u_1 responds in y_1 and input u_2 in y_2 ,
> then input $u_1 + u_2$ will respond in $y_1 + y_2$.

This can also be represented graphically; those to whom it seems an obvious and ever present property might check whether it holds true for the "simple" nonlinear function $y = u^2$.

B

An essential property of time invariant processes is that changes in output variable y do not depend on the moment in which the input

function u was given or, in other words, the response of the process is
independent of the moment in which the disturbance between time variant
and time invariant processes.

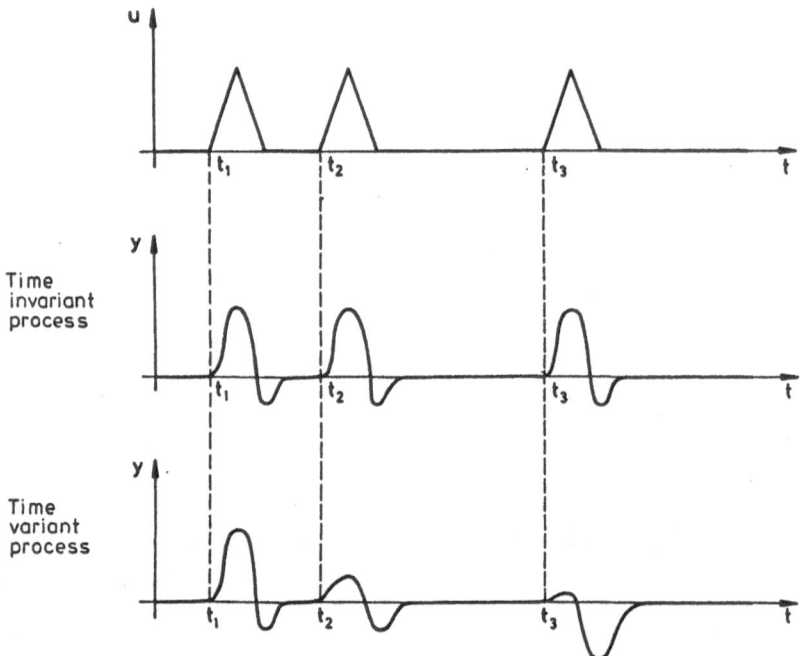

Fig. 1.1-2 An illustration of time variant and time invariant processes

It should be born in mind that, viewed through a longer period of time,
most processes show properties of time variability i.e. processes have the
tendency to change their dynamic properties. The reason for this are the
wearing, ageing, sedimentation, breakdown, and so on, of devices and
equipment in which such processes take place. (A popular example from
everyday life are changes in the dynamic properties of cars, which are as
a rule more and more sluggish from year to year.) For the needs of
control, such slow changes can be neglected and most processes can be
considered time invariant. A typical example of a time variant process is
the flight of a rocket, which burns up fuel and suffers great weight loss as
it flies. This loss of mass is no longer negligible, and must be taken into
account in a mathematical description. Consequently, the model of a
rocket flying in a straight line is

$$\frac{d(Mw)}{dt} = F(t) \; , \tag{1.1-3}$$

$$M(t) \frac{dw}{dt} + w(t) \frac{dM}{dt} = F(t). \tag{1.1-4}$$

where F is the thrust, M the rocket's mass, w the speed of flight. The time variant coefficients M(t) and w(t) are typical of time variant processes.

C

Figure 1.1-3. shows some simple examples of the differences between processes that are lumped or distributed in space.

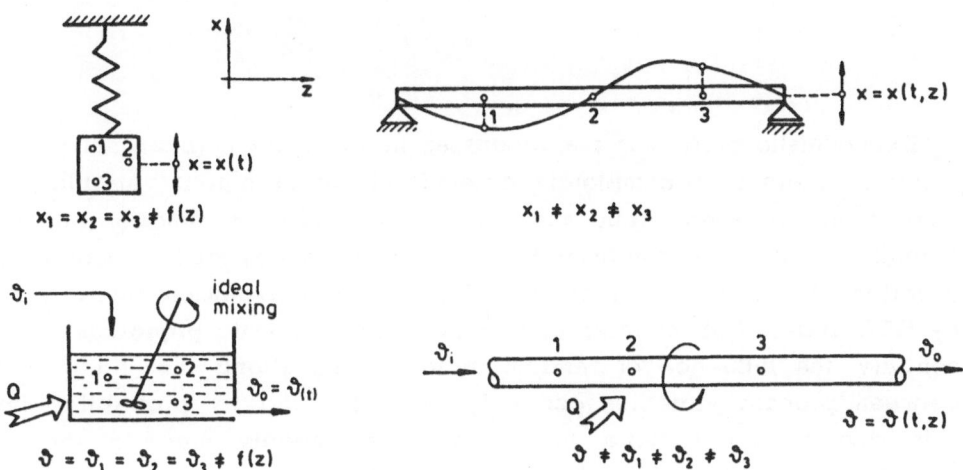

Fig. 1.1-3 An illustration of lumped and distributed processes

We will treat this point in more detail later. At present we will only briefly say that all the examples in which changes in the process variables are independent of spatial coordinates are considered to be processes with lumped parameters or processes lumped in space. Where this is not so, i.e. where some of the variables are not the same at the same moment in the whole process volume, the process is spatially distributed.

D

In flow processes mass and/or energy flow through process units in continuous currents (heat exchangers, flow chemical reactors, rolling mills, steam generators). In charge processes, on the contrary, flow is interrupted and such processes usually have three stages - charging process units, the process itself, and discharging process units (charged chemical reactors, heat processing of metal, rotational furnaces in the cement industry). In discrete part manufacturing, the objects that are to be processed appear in a discontinuous period of time (these are the majority of manufacturing processes in the metal-working and similar industries, e.g. wood, plastic and glass working).

E

Deterministic processes are processes in which for a given set of input variables there is a completely determined set of output variables. The relationship between input and output need not be unique, but it is essential for it to be determined. In their static states such processes are described by algebraic or elliptic PDE, and their dynamics are described by ODE and/or PDE or integral equations. In stochastic processes, on the contrary, the influence of random factors is so strong that relationships between process variables can only be expressed by probability laws. This division refers more to mathematical models than to the real processes they describe, because no real process can avoid random influences (noises, disturbances, breakdown). The decision on whether a process is to be considered deterministic or not depends on how much influence such random factors have on the way the process develops and on the behavior of variables of interest. Figure 1.1-4. shows typical process variable changes in deterministic and stochastic processes.

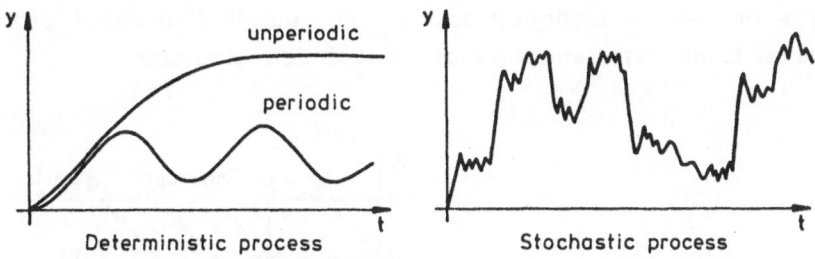

Fig. 1.1-4 Typical process variable changes

F

It has already been said that processes in which at least one process variable changes with time are considered dynamic. Thus all processes where there is no time change of state are static. In the mathematical sense this means that the time derivatives of all process variables equal zero. As a rule, a process that is already in a static regime will continue to develop under stationary conditions if all the input variables are time invariant, because the principle of causality demands a time change in the input to result in time changes in internal and output process variables also. There are, of course, exceptions. One of them is the case where process has several inputs and where changes in input variables can occur in opposite directions concerning their influence on internal and output process variables, as a result of which those variables remain unchanged.

In modern technical practice, the static properties of devices and objects in which specific processes occur are usually shown in the form of tables, characteristics diagrams, nomograms or in some other graphical manner (see for example Figs. 2.2-1.2 and 2.2-1.3 for the centrifugal pump and regulation valve). Such illustrations are obtained as the solution of one or of a system of algebraic equations or elliptical PDE, which describe the relations between process variables and are of great use in designing complex plants. However, such illustrations do not provide answers about the dynamic properties of processes that occur inside such devices. We will now give a very simple, and under certain conditions (which is truly of little importance) linear, example to illustrate the difference between a statically and a dynamically formulated problem or task, similar to the type encountered in technical practice.

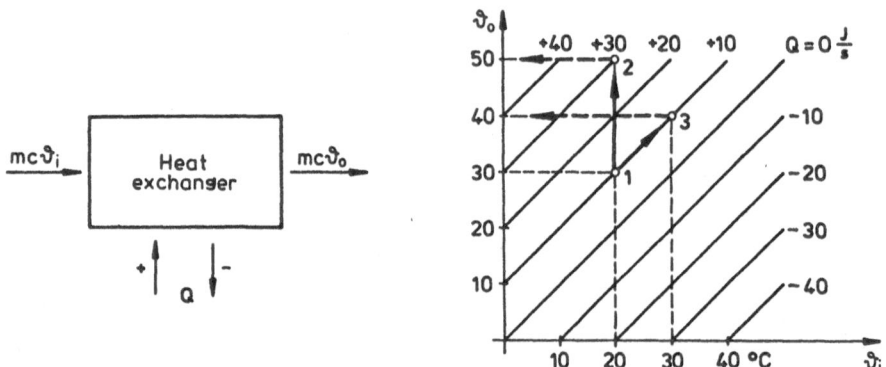

Fig. 1.1-5 Diagram of characteristics of a heat exchanger

Figure 1.1-5. shows a heat exchanger whose inner structure, flow velocity, coefficient of heat transfer, number of pipes, dimensions and the like do not interest us now. So, the equation relating output fluid temperature ϑ_0, input fluid temperature ϑ_i and the heat flow Q that is brought to the exchanger or taken from it can be written

$$mc \; \vartheta_0 = mc \; \vartheta_i \pm Q \tag{1.1-5}$$

or

$$\vartheta_0 = \vartheta_i \pm \frac{1}{mc} \; Q \; . \tag{1.1-6}$$

If flow rate m and specific heat c are constant, (1.1-6) can be easily resolved and presented in the form of a diagram of characteristics. For m = 1 kg/s and c =1 J/kgK, this diagram is a graphical representation of the functional dependence $\vartheta_0 = \vartheta(\vartheta_i, Q)$.

For a given Q = 10 J/s and ϑ_i = 20 °C, the diagram of characteristics gives one and only one point 1 for which ϑ_0 = 30 °C, which is a solution to the static problem of the heat exchanger. If input temperature ϑ_i remains constant and the heat flow increases to Q = 30 J/s, point 2 determines the temperature $\vartheta_0 = 50\ °C$, which again solves the static problem. Similarly, an increase of ϑ_i to 30 °C, keeping Q constant, determines $\vartheta_0 = 40\ °C$ as point 3.

Formulated dynamically, the problem is as follows: how long will the process take, will the temperature increase be fast or slow from 30 °C to 50 °C, and what will be the form of this change. The diagram of characteristics does not answer this question, and anyway there is an infinite number of such answers. This is because the time response ϑ_0 depends on the way in which heat flow Q changes from 10 to 30 J/s as time passes, and there is an infinite number of ways in which this can take place. Figure 1.1-6 shows only three of all the possible ways in which heat flow Q can change with time. It also shows three possible responses to such changes in heating. The solution to the dynamic problem involves determining (analytically, graphically or by numerical simulation) a time-dependent function of response $\vartheta_0(t)$ for a given $Q(t)$. This, however, cannot be done with the help of the mathematical model (1.1-6) since it reproduces only the statics of the exchanger. To solve the dynamic problem a mathematical model of dynamics, or a dynamic mathematical model must be formulated. The pages of this book are concerned with problems of obtaining dynamic mathematical models.

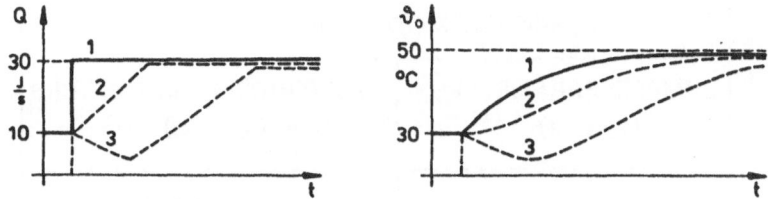

Fig. 1.1-6 Output temperature responses

Finally, we will give a survey (Table 1.1-2) of the relations between the spatial and time properties of processes and their corresponding mathematical notation and representation.

This ends the short classification of processes according to their noticeable properties. Of course, a classification could also be carried out in some other way or according to different criteria. Let that remain for narrower, more specialized works. Here we must mention that we will analyze, in the following chapters, exclusively the dynamics of both linear and nonlinear, time invariant, lumped and distributed, continuous, deterministic processes. Static (stationary) states will be of interest only inasmuch as they are necessary for specific purposes (for example, in linearization).

The solution of dynamic problems has been gaining increasingly in importance, which has made insight into the dynamic properties of processes an unavoidable and necessary part of engineering and scientific practice. When we say this, we do not mean empirical knowledge of dynamics in the sense of acquaintance with the facts and phenomena that take place inside plants in unstationary conditions (although this is useful and desirable). What we are referring to are exact mathematical expressions, equations and systems of equations, in short, mathematical models describing the dynamics we are interested in. In this last sentence, as indeed on the preceding pages also, we used the concept of a mathematical model several times. This term will also be met on the following pages. It is, therefore, necessary to say several words on the problem of the model, the way in which it is built, the forms it comes in, its properties, and so on.

Table 1.1-2

		PROCESS MODELS WITH		
		LUMPED PARAMETERS $\neq f(x, y, z)$	DISTRIBUTED PARAMETERS $= f(x, y, z)$	
T I M E	STATICS	algebric equation representations: tables nomograms characteristics	- onedimensional ODE - multidimensional eliptic PDE	$\neq f(t)$ T I M E
	DYNAMICS	- sytem of 1st-order ODE - 1 ODE of n-th order - several ODE varying order	hyperbolic PDE parabolic	$= f(t)$

SPACE

In engineering a model is considered to be a material or an idealized (symbolic) object which substitutes or represents another existing or imaginary object of interest. The basic condition that every model must fulfil is that conclusions about the properties of the original can be drawn from the behavior of the model.

On the pages of this book we are going to treat only one subgroup of all the possible groups of symbolic models - the mathematical model, about which the following can be said:

If the relations between process variables and the geometrical and physical properties of the space in which a process develops can be mapped onto a mathematical structure, then such a representation is called a mathematical model. A model can also be said to represent a symbolized hypotheses about the way in which the process under investigation will develop, and its analysis (examination and solution) gives answers about the behavior of the real process.

The model can be obtained in two ways: theoretically (also called deductively, analytically, based on the first, natural, principle) or experimentally (practically, empirically). The first procedure is called **modeling** and the second **identification**, and here are their shortest definitions:

Modeling is the process of model building using theoretical means, i.e. the fundamental natural laws of the conservation of mass, energy and momentum. **Identification** is an experimental process of model building using the measured values of input and output variables of the real process and its model. The error between process and model response should be minimised by parameter or structure change.

The pages of this book will show and treat only the theoretical method of formulating a mathematical model, but we will nevertheless present some basic features, advantages and shortcomings of the process of modeling (**M**) and the process of identification (**I**).

In **M**, model structure results directly from natural laws and possibly from some necessary neglection in the procedure of model building In **I**, structure must be predicted in advance. **M** reproduces the

relations between input, output and internal process variables while
I usually results in a "black box" type of model, i.e. a model that
reproduces only the relations between input and output. In **M**,
model parameters are directly linked with physical values and
properties and in **I**, the parameters are pure numbers, unconnected
with physical values. A model built by **M** can be used for many
operating regimes of the same process and also for related
processes about whose process variables little is often known. **I**, on
the contrary, gives a model only for the specific operating regime
for which measurements were carried out and for that particular
regime the description is reliable. In **M**, a model can be formulated
for a process that has not yet been realized in practice and is still
in design stage. Moreover, which is sometimes of particular
importance, it is possible to build a model to examine the dynamics
of breakdown processes in plants (malfunction of electrical pumps,
loss of cooling medium in nuclear reactors, burns in the wall of
a chemical reactor and the like). The **I** procedure can only be
applied to existing objects, and it is not convenient to simulate
emergency conditions on an object with the purpose of carrying out
identificational measurement. To make use of **M**, all the basic
internal processes must be known and mathematically describable.
This is not a demand for **I**. Finally, once a program package for **I**
is prepared, it can quickly and easily be applied to obtain models
for various processes. In the case of **M**, the procedure must be
started from the very beginning every time and takes up much more
time. In practice, however, both these methods very often
complement each other.

Similarly as in process classification, according to the properties of the
process under investigation, models can be either linear or nonlinear, time
variant or invariant, continuous or discrete, stable or unstable, to describe
system dynamics with lumped (then they are in the form of ODE) or with
distributed parameters (PDE).

However it is important to realize that a process or an object can have
more than one dynamic model. A **hierarchy of models in fact
exists,** from those that are dynamically the simplest, of the O-th order
(which describe only stationary states but which can be very
complicated), to very complex dynamic models of a high, and if the
process is described by PDE, of infinitely high order. Many factors
influence a model's degree of complexity, the most important being its

final objective and the technical-economic and human limitations in the procedures of its formulation and solution. This book will show how to obtain dynamic models for "pure and simple" processes that take place in objects with a simple geometry, which will make the procedure of model building somewhat shorter and more condensed than the one that generally holds for model building. This general procedure can, in principle, be divided into a sequence of stages as delineated below:

1. Problem definition, in which the objective, goals and purposes of the model are defined considering different constraints - accuracy, simplicity, human, economical and computational constraints.

2. Process or object analysis, with the purpose of determining boundaries, separating the process from its environment, defining input, output and internal state variables, dividing the process or object into simpler and elementary subprocesses or parts.

3. Assembly of conservation equations for mass (M) and/or energy (E) and/or momentum (I_m), which have the following general form:

$$\text{flow rate } (M, E, I_m)_i - \text{flow rate } (M, E, I_m)_o \pm$$

$$\pm \frac{\text{produced}}{\text{absorbed}} \text{flow rate } (M, E, I_m)_v = \frac{d(M, E, I_m)}{dt}_v \qquad (1.1\text{-}7)$$

Subscript V shows this equation to be valid for a specific and constant volume within which the accumulated M and/or E and/or I_m are considered equal and homogeneously distributed. In a lumped parameter process this is the whole process volume. In spatially distributed processes the assumption of equality holds for an infinitesimally small part of the volume dV. The third member on the left side of equation (1.1-7) has a very generalized meaning and represents, in the equation for energy conservation for example, the work incoming or outgoing from the process volume. This work is then divided into the work of the pressure force and so-called mechanical labor. If the controlled volume changes, a member describing this change must also be added to (1.1-7).

4. Assembly of physical-chemical equations of state relating process variables (ideal gas, Bernoully equation, Hooke's law). **Assembly of phenomenological equations** describing phenomena (heat transfer, Fick's law of diffusion, Arrhenius' law of chemical reactions and so on).

5. Solution and investigation of the mathematical model. Analytical solution (which is rarely possible) or simulation on computers. Nonlinear models are either simulated directly, or they are first linearized and then linear analysis is performed (modal analysis, stability, sensitivity, controllability). As a rule discretization is used to reduce PDE to a system of ODE.

6. Testing and verifying obtained results and comparing them with those that were intuitively expected. Discovering inconsistencies and unexpected model behavior (for example, if heating were to result in a decreasing fluid temperature, or the like).

7. Validation of the model - experiments are conducted with both the model and the modeled process to establish that the model predicts actual process responses to a satisfactory degree. If it is impossible to experiment on the whole object, its individual parts must be tested. If necessary, after the results have been compared, assumptions used to build the model, its structure and parameters, are changed.

8. Applying the model in accordance with the purpose it was built for.

Models in this book will as a rule be represented by a system of differential and algebraic equations. The aim of the book is to show the manners and methods of building a mathematical model of the dynamics of a process, the analysis of such models, and the relations between the real physical characteristics of a process and formalized mathematical parameters.

The following chapters, however, will, in most cases not show the procedures for solving particular differential equations or their analytical and graphical solutions. The reader is referred to the last pages of this book, to the Appendix which contains table with differential equations, transfer functions (the relation between Laplace transformations of output and input), transient functions (response, or change of output, in the case of a unit step change of input), and the zeros and poles of basic dynamic behavior - proportional, integral and derivative. In the following chapters we will classify and refer to these typical and common dynamic properties of various processes, which will soon show the need and advantage of frequent reference to that table.

The pages of the Appendix also contain a basic survey of relations between differential equations and notation in the form of matrix equations of state space, a section on problems of linearization, and pages with the basic mathematical transformations for functions that are met more often in this book. It is believed that all this may be of great use to a reader coming into contact with these problems for the first time, or after a long period of time. Therefore, before turning to the following chapter, it would be of use to look through the pages of the Appendix and renew old or gain new information, and get acquainted with the basic concepts and symbols that will be used throughout this book.

1.2 DYNAMIC VARIABLES AND COEFFICIENTS

An analysis of the dynamics of various processes reveals in all of them the existence and presence of common dynamic variables and coefficients (admittedly under still different names and completely unlike from the aspect of dimension). On the pages of this book they will be called and noted down as follows:

- dynamic variables
    accumulated (stored) variable   AV
    flow   F
    effort (potential)   E
- dynamic coefficients
    coefficient of capacity - capacitance   C
    coefficient of resistance - resistance   R
    coefficient of inertia - inertance, inductance   I

In most cases these terms and notation will mean the usual variables and coefficients, but some situations will demand that the reader accepts previously unusual terminology and concepts. This effort will, however, be rewarded by many benefits and advantages, which was why we insisted on searching for and recognizing common properties and introducing the mentioned variables and coefficients. The main reward is the possibility of perceiving dynamics integrally, as a science treating changes of state with time. The common properties of dynamic processes will manifest themselves in the equivalent forms of their mathematical models, and in responses that will show selected output variables for unit step, or some other, changes of certain input variables. According to such responses we will classify them into typical proportional, integral and derivative properties, which are given in the last table of the Appendix.

Before giving a tabular survey of the meaning of dynamic variables and coefficients, we must say something about their basic characteristics.

As a rule and according to the character of the process under observation, an accumulated variable will be the mass and/or the energy and/or the momentum (the momentum in a process with inertial properties) that exist within the space the process is developing in. (Although this book will analyse processes occurring within a space of rigid and time invariant geometry, all the discussions and models that follow can be repeated and performed under conditions of change in the control volume. In such situations, when the equations of conservation are being formulated, it is necessary to include members - which do not exist here - that also describe those changes in the process space volume.)

It will be shown that time changes of accumulated variables are the result of unbalance in mass and/or energy and/or momentum flow, and that changes of effort variables result from changes of accumulated variables directly and simultaneously. The flows themselves are either externally imposed variables, and thus independent of the processes taking place in the process space, or they are variables that depend directly on effort variables. It will be easy to check in the following chapters the scenario of causal relations and links between specific variables described here, particularly in examples where the graphical representations of variable interrelations are given in the form of block diagrams (see for example Figs. 2.1-3.3, 2.1-2.7, 2.3-2).

In this book we will always try to show mathematical models in the form of matrix equations of state space. In the case of noninertial processes it is, therefore, convenient to select state variables x so that they are accessible, measurable process variables. As a rule, an accumulated variable is not an easily and directly measurable variable, but each of its changes is simultaneously reflected in the easily measurable variables of effort. In such noninertial processes (I=0) we consequently select effort variables (pressure P, head H, concentration c, tension u, temperature ϑ and the like) as the variables of state x. In inertial processes (flow, mechanical and electrical) there is duality in selecting dynamic variables, as Table 1.2-1 shows, of which more will be said in detail in Section 2.4.

Besides the three mentioned dynamic variables which are present and noticeable in all the processes, it is also necessary to introduce into a process with inertial properties the **dynamic variable of acceleration A.** This variable is related to the variable of flow F in the following manner:

$$A = \frac{dF}{dt} , \tag{1.2-1}$$

or

$$F = \int_0^t A \, dt . \tag{1.2-2}$$

In its classical meaning the dynamic variable of acceleration A originates from mechanics, where it is written a and defined by a = dw/dt. In analogy with this, in other processes the name of acceleration has been given to the time gradient of flow F.

The concepts of coefficient of capacity C, resistance R and inertance I originate from electrotechnics, but they can also be shown to be present in other disciplines. In all the investigated processes these dynamic coefficients will be defined as follows

$$C = \frac{AV}{E} \qquad \frac{\text{accumulated variable}}{\text{effort}} \tag{1.2-3}$$

$$R = \frac{E}{F} \qquad \frac{\text{effort}}{\text{flow}} \tag{1.2-4}$$

$$I = \frac{E}{\frac{dF}{dt}} = \frac{E}{A} \qquad \frac{\text{effort}}{\text{gradient of flow}} = \frac{\text{effort}}{\text{acceleration}} \tag{1.2-5}$$

The above equations are true when the **relations** between dynamic variables AV, F and E are **linear.** In processes in which this is not so, i.e. when the relations between AV, F and E are nonlinear, C, R and I will, in a specific operating regime, be calculated from the quotients of small changes of dynamic variables around that operating regime. Expressed **mathematically, in nonlinear processes** the following is true

$$C = \frac{dAV}{dE} , \qquad\qquad (1.2\text{-}6)$$

$$R = \frac{dE}{dF} , \qquad\qquad (1.2\text{-}7)$$

$$I = \frac{dE}{d(\frac{dF}{dt})} = \frac{dE}{dA} . \qquad\qquad (1.2\text{-}8)$$

The equations (1.2-3) - (1.2-5) can also be formulated as follows.

The coefficient of capacity C equals the change in the accumulated variable required to make a unit change in potential.

The coefficient of resistance R equals the change in potential required to make a unit change in flow.

The coefficient of inertia I equals the change in potential required to make a unit change in acceleration.

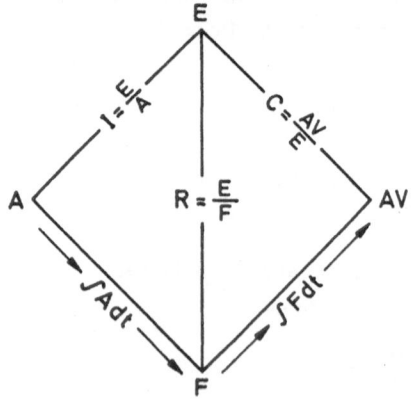

Fig. 1.2-1 Square of state or dynamic square

Everything that has been said so far shows that the relations between dynamic variables and coefficients are completely determined. This can also be represented graphically (to make them easier to remember, but not only for that reason) in what is called here the **square of state** or **dynamic square,** and is shown in Figure 1.2-1.

Dynamic variables AV, F, E and A are assigned to the corners of the dynamic square, and the dynamic coefficients I and C to its upper sides. The vertical diagonal has the coefficient R, and the bottom two sides show the relations between variables A, F and AV.

The adjective dynamic has not been used so many times in this chapter by chance. The reason is that all dynamic variables are as a rule time variant functions and they should actually be written AV(t), F(t), E(t) and A(t). Of course, theoretically those variables can be time invariant, constant, in some process. But then the problem is no longer dynamic in character and falls outside this book's field of interest. In everyday engineering practice the design, calculation and construction of various devices, process units and plants is in principle based on the assumption that those dynamic variables are truly invariant. But since it is obvious that in most cases this is not fulfilled (and especially where quality-demands for the finished product are high), a variety of measurement-regulation-automation equipment is added and built into such objects with the purpose of preserving the estimated operating conditions used in calculations in cases of always present disturbances or noises. These are the names for deviations from calculated, nominal values.

After linearization has been carried out in nonlinear processes, dynamic coefficients C, R and I are constant values and depend on the operating point in which that linearization was carried out.

Finally, we must take a look at Table 1.2-1 which shows dynamic variables and coefficients for a great many different processes, and the units used to express them. The values shown here are results, and they originate from all the following pages of this book. This table has many interesting points, but one of the things it is useful for noticing is the analogy of different physical variables. Thus, for example, the accumulated mass of gas in a tank is analogous to the amount of heat energy in the walls of a heated pipe, to the displacement of a rigid body along its path of motion, or to the momentum of a particle of liquid along a streamline.

We believe that references to Table 1.2-1 will be frequent as the chapters of this book are read. In that case we must turn the reader's attention to instances when relations between dynamic variables are nonlinear and when numerical coefficients appear as the result of linearization (which is, for example, often the case with the coefficient of resistance). Also, the variables and the coefficients we selected here are not the universally and only possible ones. For example, in the first column, AV could also have been the volume the fluid takes up in the tank, and the flow could have been the volume flow instead of the mass flow. Then C would also change and would equal area A. It is similar in the case of R and I.

If the preceding and all the following pages help and enable the reader to recognize and understand the dynamic properties of the process and the object, then this book will have fulfilled its purpose. If, therefore, the **reader is enabled** to:

- recognize the same dynamic properties in various processes,

- learn to formulate equations of conservation and, what is of special importance, develop his skill in selecting phenomena to neglect (phenomena that are of less importance for the task before him),

- foresee the consequences of assumptions, and realize how they will be transformed into the structure and the parameter values of the mathematical model,

- recognize the (usually present) nonlinear character of the process, and in connection with that accept the advantages and benefits, but also the shortcomings and limitations, of procedures of linearization,

- comprehend relations between spatially distributed processes and those that are lumped, i.e. if he learn methods for reducing models or the ways and procedures for transforming PDE into systems of ODE, to retain the knowledge and a clear feeling for what he loses and what he gains in dynamics analysis,

- relate the meanings of coefficients and the parameters of the mathematical model or, to be more precise, the eigenvalues, vectors, poles, zeros and other mathematical characteristics with the real geometrical and physical parameters of the process under observation,

- accept and get closely acquainted with the concepts of dynamic variables and coefficients, time constants, natural frequencies, proportional, integral and derivative dynamic properties, inertial and noninertial processes, and periodic and unperiodic processes, the effort that has gone into the writing of these lines will be justified.

On the following pages an analysis of spatially lumped processes begins with the simplest example of the liquid tank. Gradually, turning the pages of the book, more complex processes will be analyzed, and that **inductive path - from the simple to the more complex and from the specific to the general** - we will try to preserve within each chapter, and in the book as a whole.

Table 1.2-1 Dynamic variables and coefficients.

PROCESSES AND THEIR DYNAMIC VARIABLES AND COEFFICIENTS

	Storage – fluid	Storage – gas	Heat processes	Fluid flow (M)	Fluid flow (Iₘ=Mw)	Mech. transl. (z)	Mech. rot. (φ)	Mech. transl. (Iₘ=Mw)	Mech. rot. (L_z=Jω)	Electrical (q)	Electrical (ψ=Li)	Chemical
Direction of flow	↓	↓	↓	↓	↓	↓	↓	↓	↓	↓	↓	↓
AV accumulated variable (NL)	M	M	E	M	$I_m=Mw$	z	φ	$I_m=Mw$	$L_z=J\omega$	q	$\psi=Li$	N
units	kg	kg	J	kg	$\frac{kgm}{s}$	m	rad	$\frac{kgm}{s}$	$\frac{kgm^2}{s}$	$C=As$	$Wb=Vs$	mol
F flow	$\dot m$	$\dot m$	$\dot e$	$\dot m$	F	w	ω	F	M_z	i	u	$\dot n$
units	$\frac{kg}{s}$	$\frac{kg}{s}$	$\frac{J}{s}$	$\frac{kg}{s}$	N	$\frac{m}{s}$	$\frac{rad}{s}$	N	Nm	A	V	$\frac{mol}{s}$
E effort	H	P	ϑ	P	w	F	M_z	w	ω	u	i	c
units	m	$\frac{N}{m^2}$	K	$\frac{N}{m^2}$	$\frac{m}{s}$	N	Nm	$\frac{m}{s}$	$\frac{rad}{s}$	V	A	$\frac{mol}{m^3}$
coefficient of capacity (L, NL) $\;C=\dfrac{dAV}{dE}$; $C=\dfrac{AV}{E}$	$A\rho$	$\dfrac{V}{R_p\vartheta}$	Mc	$\dfrac{M}{P}$	$\dfrac{M}{w}$	$\dfrac{z}{F}=\dfrac{1}{c}$	$\dfrac{\varphi}{M_z}=\dfrac{1}{c_t}$	M	J	C	L	$\dfrac{N}{u}$
units	$\frac{kg}{m}$	$\frac{kg}{N/m^2}$	$\frac{J}{K}$	$\frac{kg}{N/m^2}$	kg	$\frac{m}{N}$	$\frac{rad}{Nm}$	kg	kgm^2	F	H	m^3
coefficient of resistance (L, NL) $\;R=\dfrac{dE}{dF}$; $R=\dfrac{E}{F}$	—	—	K		$\dfrac{w}{F}$	D	D_t	$\dfrac{1}{D}$	$\dfrac{1}{D_t}$	R	$\dfrac{1}{R}$	$\dfrac{u}{c}$
units	—	—	$\frac{K}{J/s}$	$\frac{N/m^2}{kg/s}$	$\frac{m/s}{N}$	$\frac{N}{m/s}$	$\frac{Nms}{rad}$	$\frac{m/s}{N}$	$\frac{rad}{Nms}$	$\Omega=\frac{V}{A}$	$\frac{A}{V}=\frac{1}{\Omega}$	$\frac{s}{m^3}$
coefficient of inertance (L, NL) $\;I=\dfrac{dE}{d(dF/dt)}$; $I=\dfrac{E}{dF/dt}$	—	—	—	$\dfrac{L}{A}$	$\dfrac{w}{dF/dt}$	M	J	$\dfrac{1}{c}$	$\dfrac{1}{c_t}$	L	C	—
units	—	—	—	$\frac{1}{m}$	$\frac{m}{N}$	kg	kgm^2	$\frac{m}{N}$	$\frac{rad}{Nm}$	H	F	—

LUMPED PROCESSES

Lumped processes are all processes in which the spatial dependence of the variables under consideration can be neglected, i.e. in which a change in those variables is considered equal and simultaneous throughout the volume for which the laws of conservation have been established. It is obvious, however, that due to the finite speed at which disturbances and variable changes propagate, all processes are in fact distributed in space. In the analysis of every individual process (and in establishing differential equations describing its dynamics) we must, thus, question how correct an analysis is if it neglects this spatial dependence. Because of the great diversity and multitude of completely different devices and plants, it is impossible to give a general criterion valid in all cases to tell us when lumped parameters are a correct substitution for the processes occurring in such objects. Nevertheless, the following definition is sufficiently wide and imprecise to serve as a guide in most cases:

Every process, in which a change in the state variable under consideration for an order-of-magnitude takes longer than it takes for the disturbance of that state variable to propagate throughout the whole process volume, can be considered a process with lumped variables.

On the following pages we will show how to build a mathematical model for the dynamics of physically different processes. We will show that the mathematical forms are similar and that the processes themselves, in spite of differences between them, all show the same or similar dynamic properties. What will be fulfilled without exception, however, is that all the models will be in the form of ordinary differential equations (ODE) and this form of presenting dynamic behavior is one of the key characteristics of all lumped processes.

At the very beginning we must stress the following. Most technical processes are nonlinear and described in the form of nonlinear ODE. The direct consequence of nonlinearity on the dynamics of the process is that its dynamic properties change depending on the operating point (on the conditions under which the process is developing). In this book, however, we will always try to linearize the differential equations and then transform such linearized equations into matrix notation in the form of state-space equations or transfer functions. This narrows down the field of values covered by the results (such models are valid only for specific operating points), but we gain very much from the possibility of applying the powerful mathematical tools developed for linear systems. In the following examples we will point out the changes in dynamic properties resulting from the nonlinearity of the process, and also show in which field of values the input variables and state variables can change, for the linearized models to still be sufficiently close to the real process.

From the aspect of process linearity and the symbols we will use in this book, the following must be said. In all the cases when the differential equations obtained are **linear,** they can also be considered **equations of the deviation of state, input and output variables** from some initial steady state. When we derive transfer functions, if we set the initial conditions equal to zero, this means that the initial variable deviations from the steady state equal zero. The transfer functions and matrix equations of state obtained in this way are valid both for absolute values of variable change from the initial state equal to zero, and also for variable deviation changes from the initial steady state. In these originally linear cases we will, therefore, not especially emphasize this linearity by introducing the deviation symbol beside the variable symbol (for example, we will not write Δm, ΔH, ΔP, ..., but just m, H, P, ...).

A. BASIC PROCESSES

2.1 MASS STORAGE

Plants often contain liquid, gas and steam tanks in which the liquid level or head H, or the gas (steam) pressure P, are of everyday interest. The technological reasons for keeping these values constant differ. Sometimes a constant H is required to keep the liquid mass fixed for the maintenance of a reaction taking place in the tank, another time this insures constant liquid supply under fixed pressure into other parts of the plant. Sometimes the tank decreases vibrations of the liquid mass in pipes (decreasing water-hammer effects) or, like in steam generator drums, insures the supply of the liquid phase into evaporating and the steam phase into superheater sections. It is similar in the case of gas (steam) tanks in which a constant pressure P insures the accumulation of a certain mass of gas (steam) or its uninterrupted supply to other parts of the plant.

Although liquids and gases are media with essentially different properties, it will be shown that the dynamics of their storage in tanks is similar, and that the physical meaning of the liquid level H is equivalent to the meaning of pressure P. Both variables show the amount of mass, liquid or gas, stored in the tank. Thus the mathematical models for both media will be obtained by formulating only one **equation for the conservation of mass.**

2.1-1 LIQUID STORAGE TANKS

Example 1 Controlled inflow and outflow of liquid

Consider the tank in Figure 2.1-1.1, into and out of which controlled amounts of liquid are pumped, $m_l \neq m_l(H)$, $m_o \neq m_o(H)$. Derive a model to describe the dynamics of the change in liquid level H. The tank is cylindrical. A = const. The compressibility of liquid is neglected. ρ = const.

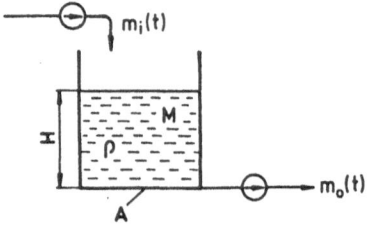

Fig. 2.1-1.1 Tank with controlled flow m_i and m_o

The equation for the conservation of liquid mass in the tank is

$$m_i - m_o = \frac{dM}{dt} = A\rho \frac{dH}{dt} \ . \qquad\qquad (2.1\text{-}1.1)$$

$$\frac{dH}{dt} = \frac{1}{A\rho} m_i - \frac{1}{A\rho} m_o \ . \qquad\qquad (2.1\text{-}1.2)$$

This is an ordinary linear DE of first order with constant coefficients.

(In these, and in many following, equations we will not stress every time that the variables H, m_i and m_o are time functions H(t), $m_i(t)$ and $m_o(t)$. Since the whole book treats dynamic processes this is always implied, and to make notation simpler this time-variability will not be specially noted.)

In the steady state, and that is the state where there is no change in head H, it must be valid that $d\bar{H}/dt=0$, which from (2.1-1.2) yields

$$\bar{m}_i = \bar{m}_o \quad . \qquad\qquad (2.1\text{-}1.3)$$

Since H does not influence the mass of liquid flowing through the tank (the functions $m_i(t)$ and $m_o(t)$ are arbitrary and externally imposed), the tank shows only properties of mass storage, and the measure for that accumulation is expressed in **capacity constant C**

$$C = A\rho \ . \qquad\qquad (2.1\text{-}1.4)$$

A more detailed explanation of the concept of capacity constant C is given in Chapter 2.1-2 in the equations (2.1-2.39) - (2.1-2.44).

The tank thus shows "pure" integral behavior. If equation (2.1-1.3) was not satisfied, the liquid level H would converge towards a theoretically infinite value (m_i > m_o) or to zero (m_o > m_i) .

All processes in which the stored value (mass, energy, momentum) does not influence inflow and outflow (mass, energy and momentum flow) from the accumulation-process volume show these integral properties.

It is not difficult to transform Equation (2.1-1.2) into a matrix DE of state space

$$[H]' = [O] [H] + \left[\frac{1}{A\rho} \quad -\frac{1}{A\rho}\right] \begin{bmatrix} m_i \\ m_o \end{bmatrix} \quad . \tag{2.1-1.5}$$

$$\mathbf{X'} = \mathbf{A} \, \mathbf{X} + \mathbf{B} \, \mathbf{U}$$

Since only one mass-storage tank is present, the system matrix **A** is of first order. As (2.1-1.5) shows, it is also a null matrix because the liquid level H has no feedback action on m_i and m_o and thus also, according to (2.1-1.1), on the rate of its own change. The eigenvalue of matrix **A** is zero

$$\lambda = 0 . \tag{2.1-1.6}$$

In proportional systems of the first order and systems with integral properties the eigenvalues of system matrix **A** have distinct physical meanings (see also Equation (2.1-1.25). Their dimension is s^{-1} , and they are related with the time constant T in the following manner

$$T = \frac{-1}{\lambda} \quad . \tag{2.1-1.7}$$

We know that the time constant T is a measure for the rate of change in the variables observed after a unit step disturbance u(t) = 1 for t > 0. In proportional systems the response reaches about 95% of its new steady state already after 3 T. For the tank with controlled m_i and m_o, Equations (2.1-1.6) and (2.1-1.7) yield an infinitely large constant T. Therefore, if Equation (2.1-1.3) is not satisfied liquid level H will reach its new steady state after an infinite period of time, or, in other words, the liquid level H will not reach a new steady state in a finite period of time. This property, that an eigenvalue lies in the origin of the s-plane, characterizes integral processes.

In the following example we will show that this tank showing integral behavior is in fact a boundary case of a tank of proportional character, in which a finite change in liquid level H does not influence the amount flowing out of the tank m_0. It will be shown that in this boundary case the resistance to outflow R becomes infinitely great.

================

Example 2 Free outflow of liquid

Formulate an equation describing unsteady changes in water level H for the tank in Figure 2.1-1.2. Inflow is controlled by a pump and outflow is free, through a control valve. All the assumptions from the preceding example are fulfilled here also, except that m_0 is no longer an imposed function, but depends on H.

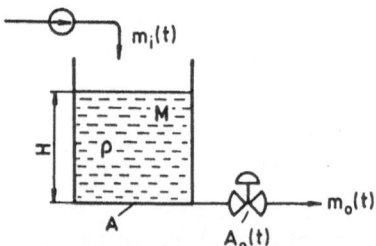

Fig. 2.1-1.2 Tank with outflow through a control valve

The equation for the conservation of mass is the same as (2.1-1.1)

$$m_i - m_0 = A\rho\frac{dH}{dt} \quad . \tag{2.1-1.9}$$

In the technically most usual case of turbulent flow through pipes, orifices, valves and fittings, we can apply the quadratic resistance law. According to this law, pressure drop is proportional to the square of flow. For outflow through a regulation valve, this law is usually shown in the following form

$$m_0 = K_V A_o \rho \sqrt{2gH} \tag{2.1-1.10}$$

The constant K_v is given for every valve by its manufacturer, and A_o is the valve's varying cross-sectional area. For a specific liquid (ρ is known), Equation (2.1-1.10) can be shown in the form of functional dependence

$$m_o = m_o(A_o,H) \qquad\qquad\qquad (2.1-1.11)$$

Substitution of (2.1-1.10) into (2.1-1.9) yields

$$A\rho\frac{dH}{dt} + K_v A_o\rho\sqrt{2gH} = m_i \qquad\qquad (2.1-1.12)$$

Equation (2.1-1.12) is a nonlinear ODE. Its nonlinearity is the result of flow through the valve and is not included only in $\sqrt{H}$ but also in the product $A_o\sqrt{H}$.

The steady state is determined by

$$\overline{m}_i = \overline{m}_o = K_v \overline{A}_o\rho\sqrt{2g\overline{H}} \qquad\qquad (2.1-1.13)$$

To linearize (2.1-1.12) we will differentiate it and replace infinitesimally small deviations df by finite deviations Δf. The nonlinear relations are shown in (2.1-1.10), whose differentiation yields

$$dm_o = \frac{\partial\overline{m}_o}{\partial A_o}dA_o + \frac{\partial\overline{m}_o}{\partial H}dH \qquad\qquad (2.1-1.14)$$

The values of the partial derivatives should be taken in the steady state in which linearization is performed. Equation (2.1-1.13) yields

$$K_v = \frac{\overline{m}_o}{\overline{A}_o\rho\sqrt{2g\overline{H}}} \qquad\qquad (2.1-1.15)$$

Referring to (2.1-1.15), after differentiation (2.1-1.13) yields

$$\frac{\partial\overline{m}_o}{\partial A_o} = \frac{\overline{m}_o}{\overline{A}_o} = K_A \; , \qquad\qquad (2.1-1.16)$$

$$\frac{\partial\overline{m}_o}{\partial H} = \frac{\overline{m}_o}{2\overline{H}} = \frac{1}{R} \qquad\qquad (2.1-1.17)$$

$$\Delta m_o = \frac{\overline{m}_o}{\overline{A}_o}\Delta A_o + \frac{\overline{m}_o}{2\overline{H}}\Delta H \qquad\qquad (2.1-1.18)$$

If we make use of these last expressions, equation (2.1-1.12) gets this linearized form for **variable deviation from the steady state**

$$A\rho \underbrace{\frac{d\Delta H}{dt}}_{C} + \underbrace{\frac{\overline{m}_o}{2\overline{H}}}_{\frac{1}{R}} \Delta H = \Delta m_i - \frac{\overline{m}_o}{A_o} \Delta A_o \qquad (2.1\text{-}1.19)$$

The resistance R is defined as the ratio of effort change to flow change. In view of (2.1-1.17), we can write

$$R = \frac{d\overline{H}}{d\overline{m}_o} = \frac{2\overline{H}}{\overline{m}_o} \qquad (2.1\text{-}1.20)$$

Considering the already introduced capacitance C, (2.1-1.19) can be written

$$T\frac{d\Delta H}{dt} + \Delta H = R\Delta m_i - RK_A\Delta A_o \quad . \qquad (2.1\text{-}1.21)$$

The constant T has the dimension of time and is thus called the **time constant.** Its analysis leads to interesting conclusions

$$T = CR = 2\frac{A\overline{H}\rho}{\overline{m}_o} = 2\frac{\overline{M}}{\overline{m}_o} = 2T_o \text{ [s]} \qquad (2.1\text{-}1.22)$$

The time constant T_o is really the ratio of the total mass stored in the tank in the steady state to the mass flow through the tank

$$T_o = \frac{\overline{M}}{\overline{m}_o} = \frac{\text{stored (accumulated) mass}}{\text{mass flow rate}} \qquad (2.1\text{-}1.23)$$

In the form of state space matrix equations, (2.1-1.21) becomes

$$[\Delta H]' = \left[-\frac{1}{T}\right][\Delta H] + \left[\frac{1}{C} \quad -\frac{K_A}{C}\right]\begin{bmatrix}\Delta m_i \\ \Delta A_o\end{bmatrix} \qquad (2.1\text{-}1.24)$$

X' = A X + B U

The eigenvalue of the system matrix **A** is obtained by solving the determinant $|A-\lambda I|=0$, whence

$$\lambda = -\frac{1}{T} \quad . \qquad (2.1\text{-}1.25)$$

Equation (2.1-1.25) confirms the already mentioned relation between the eigenvalue and the time constant in (2.1-1.7). Here it would be useful to remember and point out that the preceding example, in which the process of mass storage was integral, is really a boundary case of this example. From (2.1-1.20) we can therefore, write

$$d\overline{m}_o = \frac{dH}{R} \ .$$

(2.1-1.26)

The demand for m_o not to be a function of H really means that changes of liquid level dH will not result in changes of flow m_o, i.e. $dm_o = 0$. This is possible only if the resistance, as defined in (2.1-1.20), is infinite. From T = CR it then follows that the time constant T is infinite, and the eigenvalue of the system matrix $\mathbf{A}$ is zero.

If H is a controlled variable, i.e. a variable of interest, the output equation is

$$[\Delta H] = [1] \, [\Delta H] + [0 \ 0] \begin{bmatrix} \Delta m_i \\ \Delta A_o \end{bmatrix}$$

(2.1-1.27)

$$\mathbf{Y} = \mathbf{C} \, \mathbf{X} + \mathbf{D} \, \mathbf{U}$$

The matrix $\mathbf{D}$ is a null matrix, because there is no direct influence of the input variables on liquid level H.

The choice of sets of state or output variables is not unique and in order to show that let us now consider an example where we can vary the selection of output variables, resulting in changes of the output equation. Let the output flow Δm_o be the controlled (output) variable. The (2.1-1.14), (2.1-1.16) and (2.1-1.17) yield

$$[\Delta m_o] = \left[\frac{\overline{m}_o}{2\overline{H}} \right] \left[\Delta H \right] + \left[0 \quad \frac{\overline{m}_o}{\overline{A}_o} \right] \begin{bmatrix} \Delta m_i \\ \Delta A_o \end{bmatrix}$$

(2.1-1.28)

$$\mathbf{Y} = \mathbf{C} \, \mathbf{X} + \mathbf{D} \, \mathbf{U}$$

If there is no outflow control valve that can influence the liquid level H through changes of its outflow cross-section A_o, then we must put $\Delta A_o = 0$ in all the preceding equations. Consequently, matrices $\mathbf{B}$ and $\mathbf{D}$ and vector $\mathbf{U}$ change their dimension - the matrices lose their last columns and the vector $\mathbf{U}$ the last line.

If the Laplace transformation is applied to (2.1-1.21), using (2.1-1.29) we can show the dynamic relations between specific variables with the help of block diagrams.

$$\Delta H(s) = - \frac{1}{Ts} \Delta H(s) + \frac{R}{Ts} \Delta M_i(s) - \frac{K_A R}{Ts} \Delta A_o(s) \ . \qquad (2.1\text{-}1.29)$$

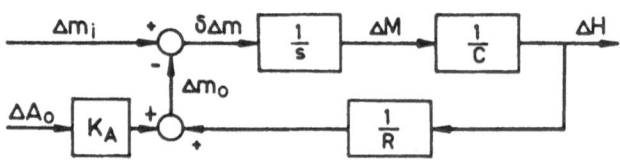

Fig. 2.1-1.3 Block diagram for a liquid storage tank

To conclude, one of the basic properties of processes that show proportional behavior (which is contrary to the properties we met in processes with integral behavior) is that the variables stored in them (mass, energy, momentum) show feedback action proportional to themselves on the flows (M, E, I_m) entering or leaving the volume observed.

Example 3 The tank under pressure

We will analyze the dynamic behavior of liquid level H for the liquid storage tank shown in Figure 2.1-1.4, which is in contact with other parts of the plant and which has constant pressure P_{in} in the air space. All the other assumptions from the former example are fulfilled here also.

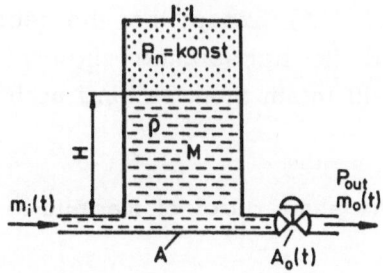

Fig. 2.1-1.4 Closed tank under constant pressure P_{in}

The law for the conservation of mass is the same as in the preceding examples, i.e. like in equations (2.1-1.1) and (2.1-1.9). Differences appear in the equation for outflow

$$m_o = K_v A_o \rho \sqrt{2gH + \frac{\delta P}{\rho}} \quad , \quad \delta P = P_{in} - P_{out} \qquad (2.1\text{-}1.30)$$

If we introduce the height of liquid H_p as an equivalent of the pressure δP

$$\frac{\delta P}{\rho} = 2gH_p \qquad (2.1\text{-}1.31)$$

equation (2.1-1.30) becomes

$$m_o = K_v A_o \rho \sqrt{2g(H + H_p)} \quad . \qquad (2.1\text{-}1.32)$$

In practice, two boundary cases are possible:

a) Very high pressure P_{in}, i.e. $H_p >> H$

Since the influence of H on the amount m_o can be neglected, Equation (2.1-1.32) now becomes

$$m_o = K_v A_o \rho \sqrt{2gH_p} \quad . \qquad (2.1\text{-}1.33)$$

Since m_o is no longer a function of H, i.e. of the mass stored in the tank, we can expect the tank to have integral properties in this case also. Truly, (2.1-1.9) and (2.1-1.33) yield

$$A\rho \frac{dH}{dt} = m_i - K_o A_o \quad , \quad K_o = K_v \rho \sqrt{2gH_p} \qquad (2.1\text{-}1.34)$$

K_0 is constant because P_{in} is kept constant in the air space. Equation (2.1-1.34) is analogous to (2.1-1.2), except that the second term on the right-hand side has changed its appearance slightly. The tank will show integral properties and will retain them as long as initial assumption $H_p >> H$ is fulfilled.

b) Pressure δP is of the same order of magnitude as liquid level H, i.e. $\delta P = 2\rho gH$.

In this case all equations in Example 2 hold, except that H^* must be inserted in all the expressions under square roots instead of H, and in the linear model ΔH^* must be inserted everywhere instead of ΔH .

$$H^* = H + H_p , \hspace{4cm} (2.1-1.35)$$

$$\Delta H^* = \Delta H \hspace{4cm} (2.1-1.36)$$

Now the tank behaves as a proportional system of the first order and its dynamics is described by Equation (2.1-1.21), i.e. by (2.1-1.24) and (2.1-1.27), except that in this case resistance R and the time constant T will be larger because H^* is larger that H.

═══════════════════════

In all the preceding examples the tank was cylindrical and placed perpendicularly on its base, so that the cross-sectional area A did not change with changes in head H. Very often tanks are spherical or conical, or cylindrical but placed horizontally (tanker trucks). In such cases the assumption A = const. is no longer fulfilled and the tank's cross-sectional area depends on the liquid level, A = A(H). The approach to analysis and the methods remain the same, except that nonlinear forms of functional dependence on liquid level now appear in equations for stored mass.

═══════════════════════

Example 4 Tank with variable cross-sectional area

Consider the horizontally placed cylindrical tank in Figure 2.1-1.5, for which an equation describing unsteady changes in liquid level H must be formulated. If the tank is filled with kerosine up to the top and the completely open valve (K_V = 0.6) has the cross-section A_o = 0.01 m^2 , find the time necessary for all the kerosine to flow out D = 2R = 2.6 m, L = 12 m.

$$r = \sqrt{H(2R-H)} \tag{2.1-1.37}$$

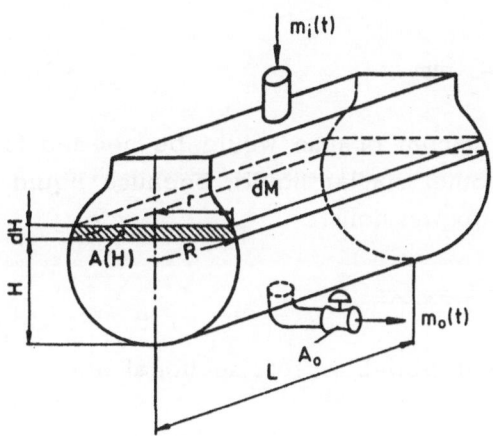

Fig. 2.1-1.6 Horizontal cylindrical tank, A = A(H)

The equation for mass conservation is in the usual form

$$m_i - m_o = \frac{dM}{dt} \tag{2.1-1.38}$$

If the liquid level changes by dH, the amount of stored mass dM is

$$dM = LA(H)\rho = L\rho 2rdH \ . \tag{2.1-1.39}$$

$$dM = 2L\rho\sqrt{H(2R-H)} \ dH \ . \tag{2.1-1.40}$$

If we substitute dM from (2.1-1.40) into (2.1-1.38) and apply Equation (2.1-1.10) for m_o, we get

$$\underbrace{2L\rho\sqrt{H(2R-H)}}_{a_1(H)} \frac{dH}{dt} + K_V A_o \rho\sqrt{2g} \ \sqrt{H} = m_i \tag{2.1-1.41}$$

This is a nonhomogeneous nonlinear ODE. Nonlinearity is contained in the square root and in the product $a_1(H)dH/dt$.

For the second part of the problem we must substitute $m_i = 0$ into Equation (2.1-1.41) and transform it into a form suitable for integration after which we get

$$\int_0^t dt = \int_0^D \frac{2L\sqrt{H(2R-H)}}{K_v A_o \sqrt{2gH}}\, dH = \frac{4LD\sqrt{D}}{3A_o K_v \sqrt{2g}}$$

$$t = 2520 \text{ s} = 42 \text{ min}$$

Note that the same amount of time would be needed for the outflow of water, alcohol or any other similar liquid, because liquid density ρ does not appear in the final expressions.

Example 5 Tank with variable cross-sectional area

Consider the tank shaped like a truncated cone and completely filled with water, shown in Figure 2.1-1.6. We must calculate the time needed for all the water to flow out of it. $D_1 = 0.8$ m, $D_2 = 0.3$ m, $H_m = 1$ m, $d = 0.03$ m, $\mu = 0.62$.

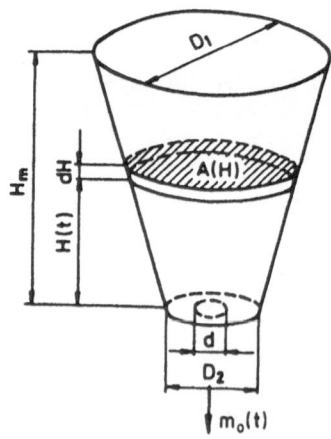

Fig. 2.1-1.6 Tank shaped like a truncated cone. $A = A(H)$

The equation for the conservation of mass for $m_i = 0$ is now

$$\frac{dM}{dt} = -m_o .$$ (2.1-1.42)

$$dM = \rho A(H)dH = \rho \frac{\pi}{4}\left[D_2 + (D_1 - D_2)\frac{H}{H_m}\right]^2 dH$$ (2.1-1.43)

If we substitute (2.1-1.43) and (2.1-1.10) (the valve coefficient K_v has been replaced by the outflow coefficient μ) into (2.1-1.42), and arrange the variables, we get

$$t = \int_0^{H_m} \frac{\rho \frac{\pi}{4}\left[D_2 + (D_1 - D_2)\frac{H}{H_m}\right]^2}{\mu \rho \frac{d^2 \pi}{4}\sqrt{2gH}} \, dH = \frac{2\sqrt{H_m}}{15d^2\mu\sqrt{2g}}(3D_1^2 + 4D_1D_2 + 8D_2)^2$$

(2.1-1.44)

$t = 194$ s $= 3$ min 14 seconds .

Similarly as in the preceding example, t does not depend on liquid density ρ here either.

―――――――――――――

The upper example of a tank shaped like a truncated cone is a typical **nonlinear case** and also a simple process in which an understanding of the basic dynamic properties does not present great difficulties. Thanks to that, on the following pages we will use this example to show one of the basic characteristics of nonlinear processes - the dependence of the dynamic properties (which are expressed in time constant T in proportional systems of the first order) on the operating (steady) state.

DEPENDENCE OF DYNAMIC PROPERTIES ON OPERATING STATE

One of the very important properties of technical devices and plants is that their dynamic properties are not the same throughout the whole range of operation. This is true of completely different processes and is independent of the system order, i.e. of the number of energy, mass and momentum storage elements. The exception are the rare processes with "purely" linear behavior. The above-mentioned dependence does not occur only in technical, but also in sociological, biological and other processes. (The example of an athlete warming up before his turn and preparing to bring his bodily and mental "system" into peak operating condition in which he can act and react faster, more dependably, strongly or precisely, is only one of the many everyday nontechnical examples of how changes in operating state affect the dynamic properties of the process.)

In proportional systems of the first order, like the ones that have appeared so far, the measure and the characteristic of those **dynamic** properties is in the first place time constant T. In this last example we will, therefore, show this time dependence on the operating (steady) state in which linearization was performed. It will be useful to bear in mind the results shown in the following lines when we analyze the dynamics of completely different or more complex processes.

For the case of $m_I = 0$ the differential equation describing nonstationary changes of liquid level in the example of Figure 2.1.-1.6 can also be written in this way

$$\underbrace{\rho A(H) \frac{dH}{dt}}_{f_1(H)} + \underbrace{\mu \rho A_o \sqrt{2gH}}_{f_2(H)} = m_I \quad . \tag{2.1-1.45}$$

Differentiating Equation (2.1-1.45), which is the way to linearization, yields

$$df_1 \frac{dH}{dt} + f_1 \frac{d(dH)}{dt} + df_2 = dm_I \quad . \tag{2.1-1.46}$$

$$df_1 = \frac{\partial f_1}{\partial H} dH \quad . \tag{2.1-1.47}$$

$$df_2 = \frac{\partial f_2}{\partial H} dH \quad . \tag{2.1-1.48}$$

$$\frac{dH}{dt} = 0 \quad . \tag{2.1-1.49}$$

Referring to the last three equations and replacing d with the finite Δ (2.1-1.46) turns into

$$\rho \bar{A}(H)\frac{d\Delta H}{dt} + \underbrace{\frac{\mu\rho A_0\sqrt{2g}}{2\sqrt{H}}}_{\dfrac{\partial \bar{l}_2}{\partial H} = \dfrac{\overline{m}_0}{2H}}\Delta H = \Delta m_l \ . \tag{2.1-1.50}$$

Arranging this expression gives

$$2\frac{\rho\bar{A}(H)H}{\overline{m}_0} \ \frac{d\Delta H}{dt} + \Delta H = \frac{2H}{\overline{m}_0}\Delta m_l \tag{2.1-1.51}$$

Time constant T is the term beside the derivative of $\Delta H'$, and if we insert the values for $\bar{A}(H)$ and $\overline{m}_0(\bar{H})$, we get

$$T = \frac{2\sqrt{\bar{H}}}{d^2 H_m^2 \ \mu\sqrt{2g}} \ \left[\bar{H}(D_1\text{-}D_2) + D_2 H_m\right]^2 \tag{2.1-1.52}$$

The last expression is clumsy, but it also clearly confirms that the time constant does truly depend on the operating state. The functional dependence of T on $\bar{H}$ is

$$T(\bar{H}) = \sqrt{\bar{H}} \ (K_0 + K_1\bar{H} + K_2\bar{H}^2) \ . \tag{2.1-1.53}$$

For the data from the preceding example $K_0 = 72.83$, $K_1 = 242.75$, $K_2 = 202.3$, and the change $T(\bar{H})$ obtained from (2.1-1.53), is shown in Figure 2.1-1.7.

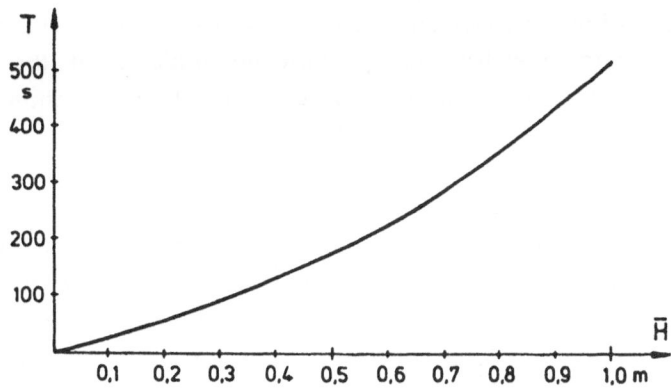

Fig. 2.1-1.7 Dependence of time constant T on operating state for the tank shaped like a truncated cone

In Example 5 (where no linearization was performed) we showed that all the liquid would run out of the tank in 194 seconds. However, if we linearize the equation describing the same outflow we get (2.1-1.51), into whose right-hand side we must put Δm_i = 0. If T is, thus, determined from the initial operating state $\bar{H}$ = 1m ($\Delta H(0)$ = 1), liquid flows out as in a proportional system of the first order, according to the exponential

$$\Delta H(t) = e^{-\frac{1}{518}t}.$$ (2.1-1.54)

As H decreases, time constant T also decreases greatly, and the error we would get from working with a constant T is more than obvious. A compromise would be to take the time constant for $\bar{H}$ = 0.5m, but even then the linear solution would not be satisfactory. Figure 2.1-1.8 shows this well. It shows the response of the original nonlinearized and of the linearized model with five different time constants calculated for $\bar{H}$ = 1, 0.75, 0.5, 0.25, 0 m.

Figure 2.1-1.8 shows the complete disproportion between the linear model response and the real response, regardless of which time constant is selected. This, however, is not unexpected. The linear model cannot be expected to cover the whole field in which the variable under consideration changes, from 0 to 100% of its initial value, especially when the process has such a pronounced nonlinearity. What is essential is that the linear and the nonlinear responses correspond quite well for the case in which linearization was performed, at the initial state $\bar{H}$ = 1 m, i.e. when T = 518 s, and for the first 10% change from the initial $\bar{H}$. In the measure of Figure 2.1-1.8, there is almost no difference in this first 10% change from the initial steady value.

The purpose of linear models is to give a sufficiently good representation of dynamic properties in the neighborhood of the operating state under observation. Most devices and plants are made to work under certain conditions. That is how they are dimensioned and their efficiency and the quality of the final products are ensured in those nominal operating regimes, which we endeavor to ensure with the help of other devices (automation, computers, regulators) or supervising personnel. Only in rare cases (starting the plant up, breakdown, shutting the plant down and the like) are changes in operating states great. In such cases, if we need models that include the dynamics of those processes, we primarily develop nonlinear models or we use linear models but change and recalculate their dynamic characteristics (parameters) after every more substantial change of operating point, to make them correspond with the operating point that has just been reached.

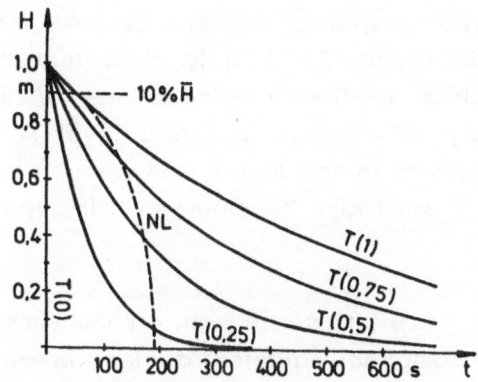

Fig. 2.1-1.8 Dynamics of change in liquid level H in the case of free outflow

----- nonlinear model response

——— linear model response with 5 different time constants

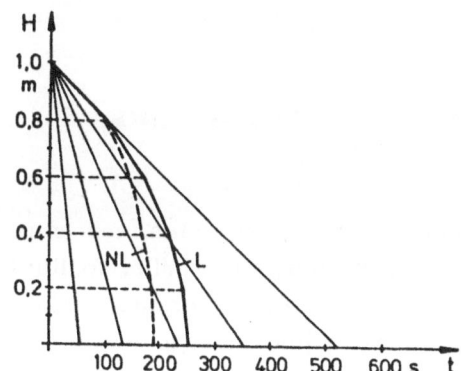

Fig. 2.1-1.9 Dynamics of change in liquid level H in the case of free outflow

----- nonlinear model response

——— linear model response with a different T for 5 operating states

The example of outflow shown in Figure 2.1-1.8 is a process with very great changes in operating state. Figure 2.1-1.9 shows the nonlinear model response and also how changes in level H would look if simulation was performed on a linear model, but if the time constant T was changed and recalculated for five different operating states ($\overline{H}$ = 1, 0.8, 0.6, 0.4, 0.2 m).

This linear-model response is an obvious improvement on the responses shown in Figure 2.1-1.8. It is clear that the "linear" response would be quite close to the "nonlinear" one if time constant T was changed after every 10% change in liquid level H. (In cases when the gain constant K appears in the linear model too, whenever the operating regime is changed K must also be changed if it depends on the operating state.)

The analysis we have just carried out on the simple process outflow is important for many completely different devices in which different forms of nonlinear relations between variables appear. Of course, the results we obtained here are not completely general in the sense that they can be applied to all cases. They are of more value as indicators of some basic relations between dynamic properties and the operating state, and also point to the different responses of the original nonlinear models and those obtained through linearization.

2.1-2 GAS AND STEAM STORAGE TANKS

Gas and steam storage tanks are standard parts of processing and energy plants. In many ways they are similar to liquid storage tanks, and gas pressure P can be considered analogous to liquid level H. The basic difference between gas and liquid storage tanks is that gases and steam are compressible so that the assumption ρ = const. is no longer valid. When the equations are formulated it is also very important to consider the conditions under which the tank is charged and discharged, and the type of gas (steam) flow. In other words, we should distinguish between subcritical and supercritical inflow and outflow, isothermal and adiabatic (or polytropic) changes in the state of gas (steam), laminar and turbulent flow.

In technical work conditions we usually have **turbulent flow** and that assumption holds throughout this whole chapter.

From the aspect of how the state of the gas (steam) in the tank changes, the following words would, in principle, be sufficient. If the tank walls are metal and uninsulated and the change slow, the gas (steam) temperature in changeable operating regimes can be considered constant. Such processes can thus be considered isothermal. In tanks that are well insulated and where heat cannot be conducted from the tank in cases of expansion, processes within the tank are adiabatic (for air the coefficient is n = 1.41). In practice, measurements have shown that the usual pressure and temperature changes in fact result as polytropic processes, somewhere between isothermal and adiabatic. For most storage tanks the coefficient n is between 1 and 1.2. In this chapter, unless stated differently, we will proceed as if gaseous (steam) changes occur **isothermally.**

Finally, when we formulate the equations we will take into account whether the tank was charged and discharged under **supercritical** or **subcritical flow conditions.**

We must also stress that we will work with the equation of state for **ideal gas.** In most cases gases and steam satisfy this equation for low pressure and for vapors that are not too close to the line of saturation. The approach and methods do not change for real gases and steam. In such cases it is only necessary to replace the equation of state for ideal gas with the equations of state for those gases and steam, and verify the results accordingly.

Finally, in this chapter also (**isothermal changes of state**), as in preceding one, models of dynamic processes will be obtained by establishing **only one equation for the conservation of mass.**

Example 1 Controlled charge and free discharge of tanks

Derive a mathematical model for the dynamic processes occurring when the state of gas (steam) in the tank in Figure 2.1-2.1 changes. Determine the equations and transfer functions for subcritical and supercritical discharge. Gas (steam) is fed into the tank from a compressor and the amount $m_i(t)$ does not depend on the pressure P. The state changes in the tank are slow enough to be considered **isothermal.**

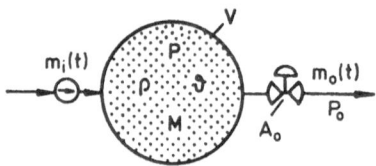

Fig. 2.1-2.1 Gas (steam) tank with controlled charge $m_i(t)$

The equation for the conservation of mass in the case of the gas in the tank is

$$m_i - m_o = \frac{dM}{dt} = \frac{d(V\rho)}{dt} = V\frac{d\rho}{dt} \quad . \tag{2.1-2.1}$$

If the purpose had been to determine changes in density ρ, then (2.1-2.1) would be the final form of the model. In practice, however, pressure P and temperature ϑ are usually regulated variables and (2.1-2.1) must be transformed to show P and ϑ. For this we will use the equation of state for gas

$$\rho = \rho(P,\vartheta) \quad , \tag{2.1-2.2}$$

from which

$$\frac{d\rho}{dt} = \frac{\partial\rho}{\partial P}\frac{dP}{dt} + \frac{\partial\rho}{\partial\vartheta}\frac{d\vartheta}{dt} \quad . \tag{2.1-2.3}$$

With (2.1-2.3), Equation (2.1-2.1) gives

$$m_i - m_o = V\frac{\partial\rho}{\partial P}\frac{dP}{dt} + V\frac{\partial\rho}{\partial\vartheta}\frac{d\vartheta}{dt} \quad . \tag{2.1-2.4}$$

Equation (2.1-2.4) contains both pressure and temperature changes and is thus generally not sufficient for a dynamic analysis. To complete the dynamic description the **equation for the conservation of energy** would also have to be formulated. However, as we have already said, gas (steam) storage processes in tanks that are not very well insulated are slow, and usually occur in isothermal conditions, ϑ = const. and $d\vartheta/dt$ = 0. This leads to

$$V\left(\frac{\partial\rho}{\partial P}\right)\frac{dP}{dt} = m_i - m_o \quad . \tag{2.1-2.5}$$

For ideal gases we have (R . . .gas constant)

$$\rho = \frac{P}{R\vartheta} \quad . \tag{2.1-2.6}$$

$$\left(\frac{\partial \rho}{\partial P}\right) = \frac{1}{R\vartheta} \quad . \tag{2.1-2.7}$$

The final result is

$$\frac{V}{R\vartheta}\frac{dP}{dt} = m_i - m_o \quad . \tag{2.1-2.8}$$

Flow rate $m_i(t)$ is determined by the work of the compressor and its dependence on P is neglected. It is thus necessary to determine $m_o(t)$, which obviously depends on whether the tank is being discharged in subcritical or supercritical conditions.

a) **Subcritical outflow**

In practice, for subcritical flow (which is defined somewhat more "roughly" by the condition $P_o > P/2$) through valves, orifices and similar parts of the pipeline, we use the formula

$$m_o = K_o A_o \sqrt{P_o(P-P_o)} \quad , \quad P_o > \frac{P}{2} \tag{2.1-2.9}$$

The exact analytical formula for calculating m_o is given in Equation (2.1-2.47), and in (2.1-2.9) K_o contains both the valve characteristic K_v and the gas characteristic in the steady state under consideration (R, ϑ, $\varkappa$) . K_v is a valve characteristic known for every valve, and A_o is the cross-sectional area through which the gas (steam) flows. In orifices or ordinary nonregulable valves A_o is constant, but in control valves $A_o = A_o(t)$, i.e. A_o is an arbitrary time function. If gas (steam) flows out the tank through an orifice, outflow characteristic μ is given instead of K_v. This characteristic includes the shape of the orifice edges, the way in which the fluid is brought to the orifice and so on. (The same is true of short parts of the pipeline.)

Substitution of (2.1-2.9) into (2.1-2.8) yields a nonlinear, nonhomogeneous ODE with variable coefficients

$$\frac{V}{R\vartheta}\frac{dP}{dt} + K_o A_o \sqrt{P_o(P-P_o)} = m_i \tag{2.1-2.10}$$

The downstream pressure $P_0(t)$ is also variable and can in practice change arbitrarily with time.

For the needs of further analysis, it will be the most useful to find the linear form of equation (2.1-2.10). Differentiation of (2.1-2.8) gives

$$\frac{V}{R\vartheta} \frac{d(dP)}{dt} = dm_i - dm_o \tag{2.1-2.11}$$

Variable dm_o is of key importance because m_o holds all the nonlinearities. If (2.1-2.9) is written in the form of functional dependence

$$m_o = m_o (P, P_o, A_o) \tag{2.1-2.12}$$

then dm_o is

$$dm_o = \frac{\partial \overline{m}_o}{\partial P} dP + \frac{\partial \overline{m}_o}{\partial P_o} dP_o + \frac{\partial \overline{m}_o}{\partial A_o} dA_o \tag{2.1-2.13}$$

In the steady state in which linearization is performed we have

$$K_o = \frac{\overline{m}_o}{\overline{A}_o \sqrt{\overline{P}_o(\overline{P}-\overline{P}_o)}} \tag{2.1-2.14}$$

Equation (2.1-2.9), together with (2.1-2.14), gives the partial derivatives of (2.1-2.13)

$$\frac{\partial \overline{m}_o}{\partial P} = \frac{\overline{m}_o}{2\delta \overline{P}} = \frac{1}{R} \quad , \quad \delta \overline{P} = \overline{P} - \overline{P}_o \tag{2.1-2.15}$$

$$\frac{\partial \overline{m}_o}{\partial P_o} = \frac{\overline{m}_o(\overline{P}-2\overline{P}_o)}{2\overline{P}_o\delta \overline{P}} = \frac{1}{R} \frac{\overline{P}-2\overline{P}_o}{\overline{P}_o} = \frac{1}{R_o} \tag{2.1-2.16}$$

$$\frac{\partial \overline{m}_o}{\partial A_o} = \frac{\overline{m}_o}{\overline{A}_o} = K_A \tag{2.1-2.17}$$

In the upper equations we have reintroduced symbols for flow resistance corresponding with the definition of resistance already given in the preceding section - see Equation (2.1-1.20). According to that definition, resistance R equals the ratio of change in effort to change in flow, and in this case of gas (steam) flow it is

$$R = \frac{\partial \overline{P}}{\partial m_o} = \frac{2\delta \overline{P}}{\overline{m}_o} \tag{2.1-2.18}$$

Figure 2.1-2.2 shows the dependence of m_o on pressure P in the tank, keeping $\overline{P}_o$ = const. The tangent slope equals the inverse value of resistance R. It is important to note that the slope of the tangent is always positive, i.e. an increase in P results in an increase in m_o.

A brief examination of Fig. 2.1-2.2 and a comparison of Equations (2.1-2.13) and (2.1-2.15) show that the first member on the right-hand side of Equation (2.1-2.13) (partial derivative $\partial \overline{m}_o / \partial P$) represents this slope of a plot of m_o versus P with P_o and A_o at their steady state values $\overline{P}_o$ and $\overline{A}_o$. The flow rate m_o drops to zero when the tank pressure P is the same as the outlet pressure P_o.

Similarly, the second partial derivative in Equation (2.1-2.13) is the slope of a plot of m_o versus P_o with P and A_o at their steady state values $\overline{P}$ and $\overline{A}_o$. This curve is represented in Figure 2.1-2.3.

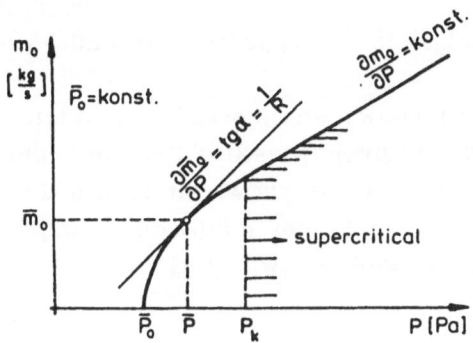

Fig. 2.1-2.2 Variation in flow rate m_o with tank pressure P, $\overline{P}_o$ = const.

Analogously to the resistance of flow to variation in the pressure P, Equation (2.1-2.16) defines resistance R_o in the case of variation in outlet pressure P_o

$$R_o = \frac{\partial \overline{P}_o}{\partial m_o} = R\,\frac{\overline{P}_o}{\overline{P} - 2\overline{P}_o} = \alpha R, \ \alpha < 0 \ . \tag{2.1-2.19}$$

It is useful to remember that in subcritical flow $(\overline{P} - 2\overline{P}_o) < 0$, therefore, the coefficient $\alpha < 0$, and consequently resistance R_o is negative. The physical meaning of this is clear - for constant P, an increase of

P_0 ($\Delta P_0 > 0$) leads to a decrease of m_0 ($\Delta m_0 < 0$) and vice versa.

This relation can be seen in Figure 2.1-2.3.

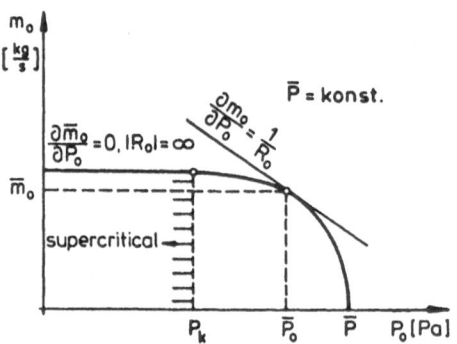

Fig. 2.1-2.3 Variation in flow rate m_0 with outlet pressure P_0 . $\bar{P}$ = const.

Equation (2.1-2.19) shows an interesting relation between resistance R_0 and pressure $\bar{P}_0$ for a given pressure $\bar{P}$ in the tank. As $\bar{P}_0$ decreases, R_0 converges to - ∞, and this is shown in Figure 2.1-2.4. When R_0 becomes infinite changes in $\bar{P}_0$ no longer influence changes in flow, i.e. $\Delta m_0 = 0$. This phenomenon is well known and occurs when supercritical flow begins.

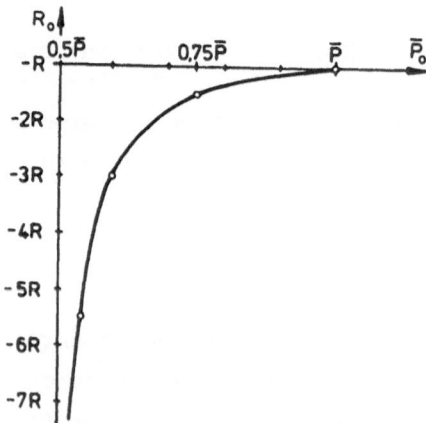

Fig. 2.1-2.4 Dependence of resistance R_0 on outlet pressure P_0.
$\bar{P}$ = const.

If the pressure difference $\delta\bar{p} = \bar{P} - \bar{P}_o$ is small (up to 10% P), we can take

$$R_o = -R = -\frac{2\delta\bar{P}}{m_o} \quad .$$

Substituting (2.1-2.15) - (2.1-2.17) into (2.1-2.13) and replacing d with the finite Δ, gives

$$\Delta m_o = \frac{1}{R}\Delta P - \frac{1}{|R_o|}\Delta P_o + K_A\Delta A_o \qquad (2.1-2.20)$$

We introduced the notation $|R_o|$ and the minus sign to stress the fact that an increase in P_o ($\Delta P_o > 0$) results in a decrease of flow rate m_o ($\Delta m_o < 0$).

Rearranging (2.1-2.11) and referring to (2.1-2.20), we get the linear form of the initial equation (2.1-2.10)

$$C\frac{d\Delta P}{dt} + \frac{1}{R}\Delta P = \Delta m_i + \frac{1}{|R_o|}\Delta P_o - K_A\Delta A_o \quad . \qquad (2.1-2.21)$$

C again denotes capacitance, and the justification of this symbol will be shown a little later. For the present, we have the equality

$$C = \frac{V}{R\vartheta} = \frac{V}{Pv} = \frac{V\bar{\rho}}{\bar{P}} = \frac{\bar{M}}{\bar{P}} \qquad (2.1-2.22)$$

Multiplying (2.1-2.21) by R we get the well-known form of an equation of a proportional system of the first order

$$T\frac{d\Delta P}{dt} + \Delta P = R\Delta m_i + \frac{R}{|R_o|}\Delta P_o - RK_A\Delta A_o \quad . \qquad (2.1-2.23)$$

If we compare this equation to Equation (2.1-2.21) for a liquid storage tank, we can see how similar the dynamic properties of those two different tanks are.

Time constant T will also show itself to be similar to the one in the preceding section. Here T is the product CR, which through the substitution of the relevant equalities becomes

$$T = CR = \frac{V}{R\vartheta}\frac{2\delta\bar{P}}{m_o} = 2\frac{\delta\bar{P}}{\bar{P}}\frac{\bar{M}}{m_o} = 2\frac{\delta\bar{P}}{\bar{P}}T^* \qquad (2.1-2.24)$$

$T^\bullet$ again represents the ratio of the mass of gas (stream) stored in the tank to the mass flow rate through the tank in the observed operating state

$$T^\bullet = \frac{\overline{M}}{m_o} = \frac{\text{stored mass}}{\text{mass flow rate}} \quad . \tag{2.1-2.25}$$

The coefficient $2\delta\overline{P}/\overline{P}$ is dimensionless.

Notation in the form of state-space equations for the case when the variables of interest are pressure deviations in the tank ΔP and deviations of flow rate through the valve Δm_o takes the form

$$\left[\Delta P\right]' = \left[-\frac{1}{T}\right]\left[\Delta P\right] + \left[\frac{1}{C} \quad \frac{R}{|R_o|}\frac{1}{T} \quad -\frac{K_A}{C}\right]\begin{bmatrix}\Delta m_i \\ \Delta P_o \\ \Delta A_o\end{bmatrix} \quad , \tag{2.1-2.26}$$

$$\mathbf{X' = A\ X + B\ U}$$

$$\begin{bmatrix}\Delta P \\ \\ \Delta m_o\end{bmatrix} = \begin{bmatrix}1 \\ \\ \dfrac{1}{R}\end{bmatrix}\left[\Delta P\right] + \begin{bmatrix}0 & 0 & 0 \\ \\ 0 & \dfrac{-1}{|R_o|} & K_A\end{bmatrix}\begin{bmatrix}\Delta m_i \\ \Delta P_o \\ \Delta A_o\end{bmatrix} \quad . \tag{2.1-2.27}$$

$$\mathbf{Y = C\ X + D\ U}$$

The process under observation, with two output (ΔP, Δm_o) and three input (Δm_i, ΔP_o, ΔA_o) variables, has 6 transfer functions. From their forms we can conclude about the dynamic properties of the process, i.e. about the way in which the output variables will vary with the input variables. The matrix transfer function $\mathbf{G}(s)$ is obtained in the usual, manner for matrix state-space equations, which we will show here in shortened form

$$\mathbf{G}(s) = \mathbf{C}(s\mathbf{I} - \mathbf{A})^{-1}\mathbf{B} + \mathbf{D} \tag{2.1-2.28}$$

$$\mathbf{G}(s) = \begin{bmatrix}1 \\ \\ \dfrac{1}{R}\end{bmatrix}\begin{bmatrix}\dfrac{1}{s + \dfrac{1}{T}}\end{bmatrix}\begin{bmatrix}\dfrac{1}{C} & \dfrac{R}{|R_o|}\dfrac{1}{T} & -\dfrac{K_A}{C}\end{bmatrix} + \begin{bmatrix}0 & 0 & 0 \\ \\ 0 & \dfrac{-1}{|R_o|} & K_A\end{bmatrix}$$

$$\tag{2.1-2.29}$$

If we perform the indicated matrix operations we get $\mathbf{G}(s)$ and we can write

$$
\begin{bmatrix} \Delta P(s) \\ \\ \Delta M_o(s) \end{bmatrix} = \begin{bmatrix} \dfrac{R}{Ts+1} & \dfrac{\frac{R}{|R_o|}}{Ts+1} & -\dfrac{RK_A}{Ts+1} \\ \\ \dfrac{1}{Ts+1} & -\dfrac{R}{|R_o|}\dfrac{Cs}{Ts+1} & \dfrac{TK_A s}{Ts+1} \end{bmatrix} \begin{bmatrix} \Delta M_i(s) \\ \Delta P_o(s) \\ \Delta A_o(s) \end{bmatrix} \qquad (2.1\text{-}2.30)
$$

$$
(T = CR, \ C = \frac{V}{R\vartheta}, \ R = \frac{2\delta\bar{P}}{m_o}, \ R_o = \alpha R)
$$

Individual transfer functions are elements of the matrix transfer function $G(s)$. They are all, except $G_{22}(s)$ and $G_{23}(s)$, transfer functions of proportional systems of the first order, which could have been expected since there is only one mass storage tank. It is especially important to note that in the transfer function $G_{12}(s)$, which relates $\Delta P(s)$ to $\Delta P_o(s)$, the gain coefficient is smaller than unity, since $R < |R_o|$. (For example, if we use Figure 2.1-2.4 we can show that if $\bar{P} = 10$ bar and $\bar{P}_o = 6$ bar, and if $\bar{P}_o$ increases to 7 bar (i.e. $\Delta P = 1$ bar), the pressure in the tank will increase only by $R/R_o = R/3$ $R = 0.333$ bar.)

Figure 2.1-2.4 also leads to the conclusion that changes in pressure deviation in the tank and in outlet flow rate will be all the smaller as $\bar{P}_o$ decreases and approaches critical pressure P_k. Since it is obvious that, when P_o converges to P_k, $|R_o|$ converges to infinity, the influence of variations in $\bar{P}_o$ will completely disappear for $\bar{P}_o \leq P_k$. In the following lines, when we analyze supercritical outflow, this will be confirmed.

The transfer functions $G_{22}(s)$ and $G_{23}(s)$ show that when ΔP_o and ΔA_o vary, outflow will respond as a derivative system with a time-lag of 1st order. This character of transfer behavior is typical for all tanks in the case of so-called downstream changes. It is not difficult to see that in the case of disturbance in ΔA_o, the flow rate change Δm_o from the preceding example of a liquid tank will also respond like derivative systems. Similar to the case of pressure change ΔP, the transfer function $G_{22}(s)$ shows that the influence of ΔP_o on the flow rate Δm_o decreases the closer $\bar{P}_o$ is to the critical pressure P_k. In the case of $\bar{P}_o \leq P_k$ this influence disappears completely.

b) **Supercritical outflow**

Often outlet pressure P_o is smaller than the critical pressure, and then gas flow through orifices is supersonic (supercritical). The basic characteristic of supercritical outflow is that the amount of gas (steam) flowing out of the tank depends on pressure P in the tank, but not on pressure P_0. It is calculated from

$$m_o = K_v A_o \psi \frac{1}{\sqrt{R\vartheta}} P \qquad . \qquad \qquad (2.1\text{-}2.31)$$

K_v is the valve characteristic, and for orifices it is the outflow coefficient μ. ψ is the gas (steam) characteristic and its value is 0.726, 0.685 and 0.669 for monoatomic, diatomic and more complex gases, respectively. If changes of state in the tank are **isothermal**, equation (2.1-2.31) becomes

$$m_o = K_o A_o P \qquad \qquad (2.1\text{-}2.32)$$

Then Equation (2.1-2.8) becomes

$$\frac{V}{R\vartheta} \frac{dP}{dt} + K_o A_o P = m_i \quad . \qquad \qquad (2.1\text{-}2.33)$$

The last equation, for A_o = const., is a linear DE with constant coefficients. For variable A_o the product $A_o P$ makes (2.1-2.33) a nonlinear equation which we can linearize to obtain

$$\frac{V}{R\vartheta} \frac{d\Delta P}{dt} + \frac{\overline{m}_o}{\overline{P}} \Delta P = \Delta m_i - \frac{\overline{m}_o}{\overline{A}_o} \Delta A_o \quad . \qquad (2.1\text{-}2.34)$$

The coefficients beside ΔP and ΔA_o are already known and represent the inverse resistance R and the coefficient of transfer K_A

$$R = \frac{\overline{P}}{\overline{m}_o} = \frac{\sqrt{R\vartheta}}{K_v A_o \psi} \quad . \qquad \qquad (2.1\text{-}2.35)$$

$$K_A = \frac{\overline{m}_o}{\overline{A}_o} \quad . \qquad \qquad (2.1\text{-}2.36)$$

Rearranging (2.1-2.34) for gas (steam) storage processes in tanks in the case of **supercritical outflow** we get

$$T\frac{d\Delta P}{dt} + \Delta P = R\Delta m_i - R K_A \Delta A_o \quad . \qquad (2.1\text{-}2.37)$$

Equation (2.1-2.37) is identical to Equation (2.1-1.21) for variable liquid level in a tank if we replace ΔH by ΔP. It differs from (2.1-2.23) for variable ΔP in the case of subsonic gas (steam) outflow because (2.1-2.37) does not contain the term showing the influence of outlet pressure deviation ΔP_0 on processes within the tank. It could also be said that the case of supersonic outflow is a limiting case of the preceding one, when P_0 converges to P_k. Thus all the equations we derived for subcritical flow are completely valid here, with two important differences:

- resistance $R = \bar{P}/\bar{m}_0$
- the middle term on the right-hand side of Equation (2.1-2.23) disappears, resulting in the disappearance of the middle column in matrices **B**, **D** and **G**(s) and the disappearance of the middle row of the input vectors **U** and **U**(s) in the matrix state-space equations (2.1-2.26) and (2.1-2.27), and in the matrix transfer functions (2.1-2.30).

Further analysis of the process is reduced to earlier comments which we need not repeat. We should, however, turn once more to the resistance R_0 to gas (steam) outflow in the case of $P_0 < P_k$, which Figure 2.1-2.3 shows to be infinite. This means that for a finite change in ΔP_0 the increase Δm_0 equals zero, i.e. that the resistance R_0 to change in flow is infinite. From the definition of resistance as the ratio of effort increase to flow increase this is even more obvious

$$|R_0| = \frac{\Delta P_0}{\Delta m_0} = \frac{\Delta P_0}{0} = \infty \quad . \tag{2.1-2.38}$$

POSSIBILITIES OF MASS STORAGE. CAPACITANCE C.

Up to now we have paid a lot of attention to the resistance certain elements of the system offer to changes in mass flow in the case of changes in effort (potential). Less has been said about possibilities of mass storage, characterized by capacitance C.

Capacitance C is defined as the ratio of increase in stored mass to increase in effort

$$C = \frac{dM}{dP} \quad . \tag{2.1-2.39}$$

In the liquid storage tank effort is the height of liquid H, and the total mass stored in the tank is $M = AH\rho$. From (2.1-2.39) we get

$$C = \frac{d(AH\rho)}{dH} = A\rho = \frac{M}{H} \cdot \left[\frac{kg}{m}\right] \quad . \tag{2.1-2.40}$$

In the gas (steam) storage tank the equivalent of height H is pressure P in the tank, but in this case the possibility of gas (steam) accumulation also depends on the conditions under which that fluid is stored. From that aspect we distinguish between isothermal C and polytropic C_n capacitance C. As has already been said, in this text we will usually analyze isothermal processes and use the ideal gas equation. Therefore

$$C = \frac{dM}{dP} = \frac{d(V\rho)}{dP} = \frac{V}{R\vartheta} = \frac{M}{P} \cdot \left[\frac{kg}{Pa}\right] \tag{2.1-2.41}$$

In the case of polytropic temperature change ϑ is no longer independent of pressure P so we have

$$Pv^n = const. \quad , \tag{2.1-2.42}$$

that is

$$\frac{dv}{dP} = -\frac{1}{n}\frac{v}{P} \quad . \tag{2.1-2.43}$$

Consequently, the polytropic capacitance C_n is obtained as follows

$$C_n = \frac{dM}{dP} = V\frac{d\rho}{dP} = -\frac{V}{v^2}\frac{dv}{dP} = \frac{1}{n}\frac{V}{v}\frac{1}{P} = \frac{1}{n}C \quad . \tag{2.1-2.44}$$

Thus capacitance C is the greatest in the case of isothermal changes of state in gas (steam).

Example 2. Outflow of gas. Calculation of time constant T

Consider a tank of volume $V = 50$ m^3 filled with gas at the temperature of $\vartheta = 293$ °C with a gas constant $R = 150$ J/kgK and under the pressure $P = 10$ bar ($x = 1.4$, $\psi = 0.685$). At the tank exit is a valve of area $A_o = 0.002$ m^2 with valve characteristic $K_v = 0.75$. Determine time constant T that characterizes the dynamics of pressure change in the tank if the outflow is into:

a) the atmosphere,
b) into a tank with constant pressure $P_o = 6$ bar.

a) The critical pressure for a pressure of $P = 10$ bar in the tank is $P_k = 5.28$ bar. Since $P_o = 1$ bar, outflow is supercritical. From (2.1-2.31) it follows that

$$\frac{dm_o}{dP} = K_v A_o \psi \frac{1}{\sqrt{R\vartheta}} = \frac{1}{R} \qquad . \qquad (2.1\text{-}2.45)$$

Assuming an isothermal change in gas state in the tank, we have (2.1-2.41), which together with (2.1-2.45) determines T

$$T = CR = \frac{V}{R\vartheta} \frac{\sqrt{R\vartheta}}{K_v A_o \psi} = 232.1 \text{ s.} \qquad (2.1\text{-}2.46)$$

b) In the case of $P_o = 6$ bar flow is subcritical so time constant T can be determined from (2.1-2.24).

$$T = CR = \frac{V}{R\vartheta} \frac{2\delta\bar{P}}{\bar{m}_o} \qquad .$$

In the upper equation the only unknown variable is flow rate m_o, which is determined from (2.1-2.9) for subcritical flow. But in this case we will use the more complex and correct mathematical expression

$$\bar{m}_o = K_v \bar{A}_o \bar{P} \sqrt{2\frac{\varkappa}{\varkappa-1}\frac{1}{R\vartheta}\left[(\frac{\bar{P}_o}{\bar{P}})^{\frac{2}{\varkappa}} - (\frac{\bar{P}_o}{\bar{P}})^{\frac{\varkappa+1}{\varkappa}}\right]} \qquad . \qquad (2.1\text{-}2.47)$$

Equation (2.1-2.9), which we used until now, is satisfactory for most practical calculations, and constant K_o can, after the flow rate $\bar{m}_o$ has been calculated from (2.1-2.47), be obtained easily from (2.1-2.14). The upper equation gives $\bar{m}_o = 4.84$ kg/s, from which it is not difficult to get

$$T = 188.04 \text{ s.}$$

Commenting on Equation (2.1-2.33) we said that if the valve area A_o is constant, the discharge of the tank under supercritical conditions is linear if gas-state changes in the tank are isothermal. (In that case temperature ϑ beside the dP/dt term is constant.) The following example will show that it is then relatively easy to describe the dynamics of pressure change in the tank.

Example 3. Outflow of air under supercritical conditions

Air flows into the atmosphere out of a tank of volume $V = 400$ dm^3, with a pressure $P = 40$ bar and temperature $\vartheta = 15$ °C, through an orifice $A_o = 20$ mm^2 and $\mu = 1$. $P_o = 1$ bar. Determine how long it will take to reach subcritical outflow. We assume that the process in the tank is slow enough to be considered isothermal. For air, $\psi = 0.685$ and $R = 287$ J/kgK.

For $P_o = 1$ bar, $P_k = P_o/0.528 = 1.894$ bar. Equation (2.1-2.33) with $m_i = 0$ gives

$$T\frac{dP}{dt} + P = 0 \qquad . \tag{2.1-2.48}$$

Equation (2.1-2.46) determines time constant T.

$$T = \frac{V}{\sqrt{R\vartheta}}\frac{1}{\mu A_o \psi} = 101.54 \quad s \qquad .$$

The solution of (2.1-2.48) is simple

$$P = 40\ e^{-\frac{1}{T}t} = 40\ e^{-0.00985\,t} \qquad .$$

The time after which P will equal 1.894 is obtained from

$$t_{krit} = \frac{1}{0.00985}\ \ln\frac{40}{1.894} = 310\ s \qquad . \tag{2.1-2.49}$$

The assumption of temperature constancy in the tank can be made for "slow" processes of state change. Many tanks are charged and discharged so quickly that heat cannot be completely exchanged with the surroundings in that short time and the process in the tank is polytropic. The purpose of the following example is to show the error if we use an isothermal model for a process that is in fact polytropic. (Of course, in every particular case a similar procedure should be repeated because of testing.)

Example 4. Linearity of gas (steam) outflow from tank

Consider an air tank of volume V = 60 dm³ for starting up Diesel aggregate. The air pressure is 60 bar and the temperature 293 °C. As the aggregate is started up, the air flows through an orifice 5 mm in diameter into a cylinder where the pressure is 1 bar. The outflow coefficient is μ = 0.6, and the air pressure in the tank decreases polytropically with n = 1.25. For air we have ψ = 0.685, R = 287 J/kgK. Determine the time necessary for the pressure in the tank to decrease to 12 bar and show the pressure change graphically for a:

a) polytropic process,
b) isothermal process.

a) For P_0 = 1 bar, P_k = 1.894 bar and the flow conditions are supercritical all the time. For m_i = 0, using (2.1-2.1) and (2.1-2.44), we can write

$$\frac{dM}{dt} = \frac{dM}{dP}\frac{dP}{dt} = C_n\frac{dP}{dt} = -m_0 \quad . \tag{2.1-2.50}$$

Since outflow occurs under supercritical conditions, we take (2.1-2.31) for m_0 and get

$$-\frac{V}{nR\vartheta}\,dP = \mu A_0\psi\,\frac{P}{\sqrt{R\vartheta}}\,dt \tag{2.1-2.51}$$

Using the known relation $\vartheta = \vartheta_1(\frac{P}{P_1})^{\frac{n-1}{n}}$, after arranging (2.1-2.51)

we see that

$$dt = -\frac{V}{n\mu A_0\psi\sqrt{R\vartheta_1}}\,\frac{P_1^{\frac{n-1}{2n}}}{P^{\frac{3n-1}{2n}}}\,dP \tag{2.1-2.52}$$

Integration of the upper equation from P_1 to P, where P_1 is the initial pressure (in correspondence with the notation we used up to now, we could write $\bar{P}$ instead of P_1) and P the pressure at t, we get the final result for time t

$$t = \frac{V}{\mu A_o \psi \sqrt{R \vartheta_1}} \frac{2}{n-1} \left[(\frac{P_1}{P})^{\frac{n-1}{2n}} - 1 \right] \qquad (2.1-2.53)$$

Thus, the time needed for the pressure in the tank to fall from 60 to 12 bar, if the change is polytropic, is from (2.1-2.53)

$$t = 36 \text{ s.}$$

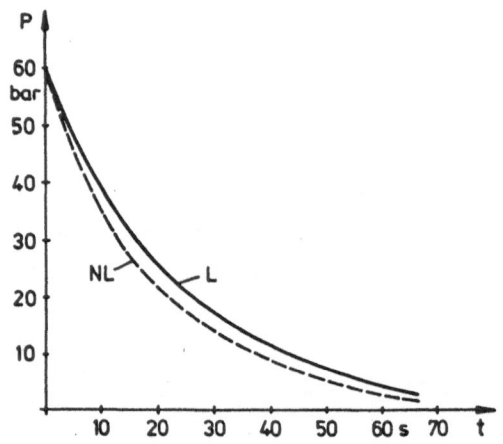

Fig. 2.1-2.5 Pressure decrease in the tank
----- polytropic (real) change. NL.
——— isothermal change. L.

b) As in the preceding example, we have

$$T \frac{dP}{dt} + P = 0 \quad ,$$

where

$$T = CR = \frac{V}{R \vartheta_1} \frac{\sqrt{R \vartheta_1}}{\mu A_o \psi} \approx 25.64 \text{ s} \quad .$$

For the initial condition $P_o = P_1 = 60$ bar, the solution of the upper DE is

$$P = 60 \ e^{-\frac{1}{T}t} = 60 \ e^{-0.039t} \quad . \qquad (2.1-2.54)$$

The last equation for $P = 12$ bar yields.

$$t = \frac{1}{0.039} \ln \frac{60}{12} = 41.27 \text{ s} \quad .$$

The error in the time necessary for pressure in the tank to fall from 60 to 12 bar, if the change is considered isothermal, is +14.64%. Figure 2.1-2.5 shows the curves of the real polytropic (nonlinear - broken line) and of the isothermal (linear - full line) pressure fall obtained from (2.1-2.53) and (2.1-2.54).

These curves show that the error due to the assumption of isothermal state change in the tank is not very great. However, the results given in this example are not generally valid. Their main purpose is to serve as an illustration for the possible consequences of some assumptions.

━━━━━━━━━━

To end the preceding examples, we must repeat that when the tank is discharged under supercritical conditions, the process is linear only if an isothermal state change takes place within the tank. In that case the time constant T is unchanged in every moment, since the temperature ϑ is constant - see Equation (2.1-2.46).

━━━━━━━━━━

Example 5. Charge and discharge of tank due to pressure difference

Consider the gas (steam) tank shown on Figure 2.1-2.6, which is charged and discharged through an inlet and outlet (control) valve due to pressure difference. We must formulate the equations describing unsteady changes of gas (steam) pressure in the tank depending on pressures P_i and P_o, and on the area of the control valve A_o. The changes in the tank are isothermal. A_i = const.

Unlike Example 1, the amount of gas at the tank inlet is no longer independent of the pressure P. Analogously to Equation (2.1-2.9), i.e. its linear form (2.1-2.20), in the case of **subcritical inflow** we see that

$$m_i = K_i A_i \sqrt{P(P_i - P)} , \qquad P > \frac{P_i}{2} , \qquad (2.1-2.55)$$

$$\Delta m_i = \frac{1}{R_i} \Delta P_i - \frac{1}{|R^{\bullet}|} \Delta P \quad , \tag{2.1-2.56}$$

$$R_i = \frac{\partial \bar{P}_i}{\partial m_i} = \frac{2\delta \bar{P}_i}{\bar{m}_i} \quad , \quad \delta \bar{P}_i = \bar{P}_i - \bar{P} \quad , \tag{2.1-2.57}$$

$$R^{\bullet} = \frac{\partial \bar{P}}{\partial m_i} = R_i \underbrace{\frac{\bar{P}}{\bar{P}_i - 2\bar{P}}}_{\alpha_i} = \alpha_i R_i \quad , \quad \alpha_i < 0. \tag{2.1-2.58}$$

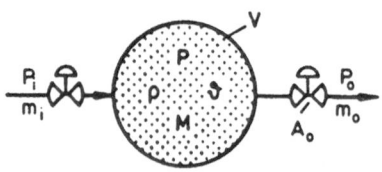

Fig. 2.1-2.6 Gas (steam) tank with uncontrolled inflow $m_i(t)$

For small pressure differences between P_i and P, α_i is close to unity and we have $|R^{\bullet}| = |R_i|$. The same diagram from Figure 2.1-2.4 can be used for α_i, only now $R^{\bullet}$ is on the ordinate and pressure P, expressed in parts of the input pressure P_i, on the abscissa. All the other equations and methods used in Example 1 are valid here also, and need not be repeated. For a given $m_i(t)$, Equation (2.1-2.21) yields

$$C \frac{d\Delta P}{dt} + \frac{1}{R} \Delta P - \frac{1}{|R_o|} \Delta P_o = \frac{1}{R_i} \Delta P_i - \frac{1}{|R^{\bullet}|} \Delta P - K_A \Delta A_o \tag{2.1-2.59}$$

If an equivalent resistance R_e is introduced in the following manner

$$\frac{1}{R_e} = \frac{1}{R} + \frac{1}{|R^{\bullet}|} \quad . \tag{2.1-2.60}$$

Equation (2.1-2.59) is translated into a form very similar to (2.1-2.23)

$$T \frac{d\Delta P}{dt} + \Delta P = \frac{R_e}{R_i} \Delta P_i + \frac{R_e}{|R_o|} \Delta P_o - R_e K_A \Delta A_o \tag{2.1-2.61}$$

$$T = C R_e = \frac{C}{\dfrac{1}{R} + \dfrac{1}{|R^{\bullet}|}} \quad . \tag{2.1-2.62}$$

We will not rewrite the state-space equations from Example 1. The Equations (2.1-2.26), (2.1-2.27) and (2.1-2.30) given there are completely valid here also, with the following changes

$$R = R_e \quad ,$$

$$T = CR_e = \frac{\dfrac{V}{R\vartheta}}{\dfrac{1}{R} + \dfrac{1}{|R^{\bullet}|}} \quad .$$

$$\Delta m_i = \frac{\Delta P_i}{R_i} \quad , \quad \Delta M_i(s) = \frac{\Delta P_i(s)}{R_i} \quad .$$

We will not specially analyze cases when charging and discharging occur under supercritical conditions either. The upper equations are general enough and can also be used in supercritical cases, when it is sufficient to make the resistance infinite. For example, if the tank is charged under supercritical conditions, i.e. for $P < P_i/2$, the coefficient $R^{\bullet} =. \infty$, $R_e = R$, and R_i also changes, $R_i = \bar{P}_i/\bar{m}_i$, so that (2.1-2.61) becomes

$$T\frac{d\Delta P}{dt} + \Delta P = \frac{R}{R_i}\Delta P_i + \frac{R}{|R_o|}\Delta P_o - RK_A\Delta A_o \qquad (2.1\text{-}2.63)$$

$$T = CR$$

Equation (2.1-2.61) shows that now time constant T is not determined only by coefficient R, and according to (2.1-2.18) by the pressures $\bar{P}$ and $\bar{P}_o$, but also by $R^{\bullet}$ and consequently, also by pressure $\bar{P}_i$. This dependence on P_i ceases to exist only when the tank is charged under supercritical conditions. Also, if ΔP_i or ΔP_o change, ΔP will not change for the same value but its change is determined by the gain coefficients $K_{pi} = R_e/R_i$ and $K_{po} = R_e/|R_o|$.

At the end of this example it will be of use to show the interdependence of variables from Equation (2.1-2.61) in a block diagram. We do this not only because a graphical representation is convincing, but also because the diagram imposes comparison between the dynamics of heat transfer through the wall of the heat exchanger and these processes of mass storage. The structure of the blocks in Figures 2.1-2.7 and 2.3-2 is completely identical. The only small difference is that in Figure 2.1-2.7 the resistances in the feedback link differ from the resistance beside the input variables. (If the differences between pressures P_i, P and P_o are not great, those resistances become the same, i.e. we will have $R_o = R$ and $R^* = R_i$, so the diagrams become identical.)

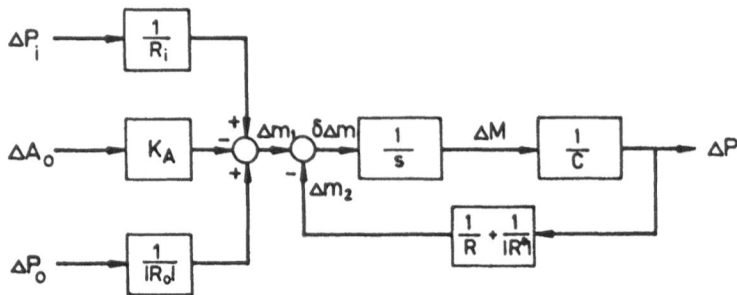

Fig. 2.1-2.7 Block diagram of pressure dynamics in a gas (steam) tank

In the previous discussion of tanks that could only store mass, the mathematical model of dynamics was obtained from the equation for the conservation of mass. In the following section we will show how the same or similar dynamic phenomena and their notation appear in the physically completely different processes of fluid flow through pipes. Their models will either be obtained only from the **equation for the conservation of momentum** or from the **simultaneous** formulation of the **equation for the conservation of momentum** and the **equation for the conservation of mass.**

2.2 FLUID FLOW

Fluid supply and transport and the connection of individual process units runs through pipes (steam pipes, gas pipes, waterpipes, pipelines), which together with pumps, compressors, valves and other fittings are the unavoidable composite parts of plants. The process of fluid flow rates through pipes is usually a controlled process in the sense that pressures and flows are kept within predetermined boundaries through the action of control devices on various executive control elements or actuators (usually regulation valves and pumps or compressors). It is thus also important to know the dynamic properties of those flow processes. The basic equations describing flow processes are identical for liquids, gases and steam. Nevertheless, for practical needs, to simplify the often complicated calculations that must be undertaken if we use the most general laws and equations of fluid mechanics, it is useful to analyze liquid flow separately from gas and steam flow. In the first case we can usually neglect the compressibility of liquids (ρ = const.) but not momentum, i.e. the inertia of liquid in the pipe. In the case of gases and steam the procedure will be the opposite. It will usually be possible to neglect their momentum, but not also the effects of compressibility, i.e. we will have to count on changes in density ρ in unsteady conditions.

On the following pages, when we devise models for the dynamics of flow processes, we will start from simpler and more specific cases of liquid and gas (steam) flow making use of the upper hypotheses. Gradually we will reach the most general dynamic model for the flow of an arbitrary fluid that has both the property of capacitance ($\rho \neq$ const., C $\neq$ 0) and the property of inertia (I $\neq$ 0).

2.2-1 LIQUID FLOW

The basic assumption in this part of the book is that liquid is incompressible (ρ = const.), which in fact means neglecting liquid storage in pipes, i.e. we consider that capacitance C = 0. In this way, at the very beginning, we dismiss the possibility of observing the well-known water hammer effects in pipes. In Section 2.2-3 we will leave out this assumption and get a model for the flow of an arbitrary fluid, for which the models in this chapter (C = 0) are only a special case. Unlike in the preceding section, here we neglected compressibility so we will now obtain a model by formulating an **equation for the conservation of momentum.**

Example 1 Dynamics of liquid flow through pipes

Figure 2.2-1.1 shows a horizontal pipe of length L, with solid walls and of constant cross-sectional area A, which connects two tanks. A centrifugal pump pumps liquid through the pipe and a control valve in the pipe regulates flow. Derive a model for unsteady flow changes through the pipe. The liquid is considered incompressible (ρ = const., m $\neq$ m(z)).

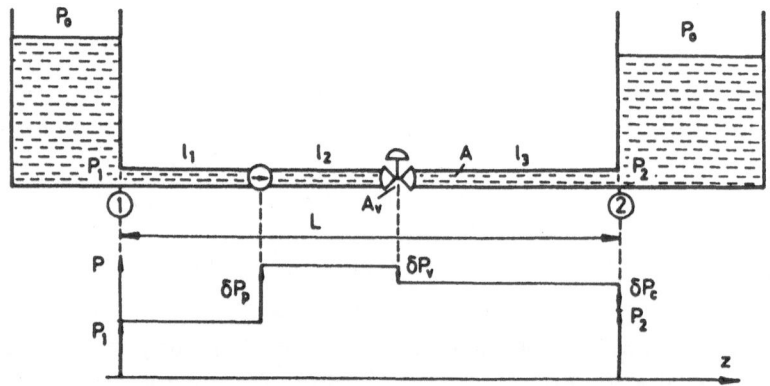

Fig. 2.2-1.1 Liquid transport between two tanks

The law for the conservation of momentum (I_m = M w) for liquid in a pipe has the well-known form

$$\sum_{i=1}^{n} F_i = \frac{dI_m}{dt} \qquad\qquad (2.2\text{-}1.1)$$

F_i denotes all the forces that act on the total mass (M = ALρ) of liquid in the pipe. In the general unsteady case pressures P_1 and P_2, pressure changes in the valve (drop δP_v) in the pipe (drop, decrease δP_c) and in the pump (increase δP_p) are time functions. If we specify the forces that act on the mass of liquid M in Equation (2.2-1.1), we get

$$(P_1 + \delta P_p - \delta P_v - \delta P_c)A - P_2 A = \frac{d(Mw)}{dt} = AL\rho\frac{dw}{dt} \; . \qquad (2.2\text{-}1.2)$$

The liquid flow rate (mass flow m) is

$$m = Aw\rho \; . \qquad\qquad (2.2\text{-}1.3)$$

and if we insert velocity w from the upper equation into Equation (2.2-1.2), we get

$$P_1 + \delta P_p - \delta P_v - \delta P_c - P_2 = \frac{L}{A}\frac{dm}{dt} \quad . \tag{2.2-1.4}$$

This equation describes the change of mass flow through the pipe depending on pressures P_1 and P_2 and pressure changes δP_p, δP_v and δP_c. The greatest mathematical difficulties in obtaining models hide in the determination of those pressure decreases and increases. The same result can be found in Section 3.3, when the fluid flow is observed as a process with distributed parameters (see Equation (3.3-29)).

It is a known fact, and in the case of valves it was shown in the preceding section, that there is a direct relationship between flow rate m and pressure change δP in individual parts of the fittings. For pumps and valves these changes depend on many other factors as well. In the case of pumps the speed of rotation n is of basic importance, and for valves their cross-sectional area A_v. In short, if we want to get a final expression for dynamic flow changes, we must know the following analytical relations between flow and pressure change

$$\delta P_p = \delta P_p(m, n) \quad . \tag{2.2-1.5}$$

$$\delta P_v = \delta P_v(m, A_v) \quad . \tag{2.2-1.6}$$

$$\delta P_c = \delta P_c(m) \quad . \tag{2.2-1.7}$$

As a rule the upper analytical expressions are not available, but there is (always supplied with the equipment) a graphical presentation of pump and valve characteristics. Figures 2.2-1.2 and 2.2-1.3 show those characteristics, and they also show how to determine graphically the values of the partial derivatives of pressure δP_p and δP_v by the independent variables.

For turbulent flow through pipes we have the characteristic resistance law, according to which

$$\delta P_c = \lambda\frac{L}{d}\frac{w^2}{2}\rho = \lambda\frac{L}{d}\frac{m^2}{2A^2\rho} = K_c m^2 \quad . \tag{2.2-1.8}$$

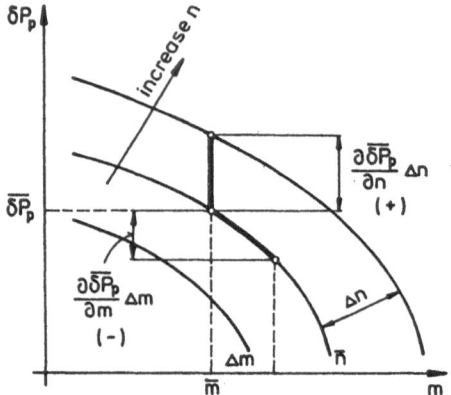

Fig. 2.2-1.2 Static characteristics for a centrifugal pump

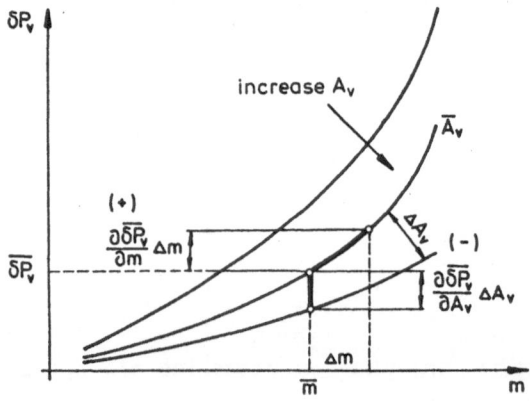

Fig. 2.2-1.3 Static characteristics for a regulation valve

The curves shown and Equation (2.2-1.8) make it obvious that Equation (2.2-1.4) is nonlinear, and if we want to analyze unsteady flow it should be linearized. This narrows down the field of process variable changes for which the model gives a faithful dynamics description, but linearized models make it simpler to understand the character of transient processes, to carry out analyses and to compare the dynamics of flow processes with other, completely different, processes.

Differentiating Equations (2.2-1.4) - (2.2-1.7) we get

$$dP_1 + d\delta P_p - d\delta P_v - d\delta P_c - dP_2 = \frac{L}{A} \frac{d(dm)}{dt} \; . \qquad (2.2\text{-}1.9)$$

$$d\delta P_p = \underbrace{\frac{\partial \delta \overline{P}_p}{\partial m}}_{R_p} dm + \underbrace{\frac{\partial \delta \overline{P}_p}{\partial n}}_{b_p} dn \; , \qquad (2.2\text{-}1.10)$$

$$d\delta P_v = \underbrace{\frac{\partial \delta \overline{P}_v}{\partial m}}_{R_v} dm + \underbrace{\frac{\partial \delta \overline{P}_v}{\partial A_v}}_{b_v} dA_v \; , \qquad (2.2\text{-}1.11)$$

$$d\delta P_c = \underbrace{\frac{\partial \delta \overline{P}_c}{\partial m}}_{R_c} dm \; . \qquad (2.2\text{-}1.12)$$

In the last three equations, in correspondence with the already given definition for resistance to flow R, we introduced symbols for pump resistance R_p, valve resistance R_v and pipe resistance R_c. If analytical expressions in the form of Equations (2.2-1.5) - (2.2-1.7) exist, it is in most cases not difficult to calculate coefficients R and b. If there are no such expressions, however, coefficients R and b in the steady state are determined graphically, as the preceding figures show.

On the basis of previous analysis, we can write without repeating the derivation

$$R_v = \frac{2\delta \overline{P}_v}{m} \; , \qquad (2.2\text{-}1.13)$$

$$b_v = -\frac{2\delta \overline{P}_v}{A_v} \; , \qquad (2.2\text{-}1.14)$$

$$R_c = \frac{2\delta \overline{P}_c}{m} = 2\overline{K}_c \overline{m} \; . \qquad (2.2\text{-}1.15)$$

It is important to point out that as flow through the pump increases, the pump head decreases. Thus resistance R_p is negative. Finally, we can write

$$\Delta \delta P_p = -|R_p|\Delta m + b_p \Delta n \; , \qquad (2.2\text{-}1.16)$$

$$\Delta \delta P_v = R_v \Delta m - |b_v|\Delta A_v \; . \qquad (2.2\text{-}1.17)$$

$$\Delta \delta P_c = R_c \Delta m \quad . \tag{2.2-1.18}$$

If we substitute the last three equations into (2.2-1.9), we get

$$\frac{L}{A} \frac{d\Delta m}{dt} + (R_v + R_c + |R_p|)\Delta m = \Delta P_1 + b_p \Delta n + |b_v|\Delta A_v - \Delta P_2 \tag{2.2-1.19}$$

$$T \frac{d\Delta m}{dt} + \Delta m = \frac{1}{R} \Delta P_1 - \frac{1}{R} \Delta P_2 + \frac{b_p}{R} \Delta n + \frac{|b_v|}{R} \Delta A_v \tag{2.2-1.20}$$

The upper equation shows that unsteady changes of mass flow of an incompressible liquid through a pipe have the properties of the first-order (proportional) system. This is a general equation with four input variables on the right-hand side. In practice, if the valve cannot be regulated the term with ΔA_v on the right-hand side disappears. Similarly, if the pump rotates with a constant number of revolutions the term with Δn is lost, and so on.

It is especially important to analyze the time constant T. We must know how it came into being and what its value is. Before that, however, we must get acquainted with the concept of inertance I of the mass of liquid.

Let liquid flow ideally, without friction, through a smooth pipe of length L and cross-sectional area A. Then $M = AL\rho$ denotes the total mass of liquid in the pipe, and $m = Aw\rho$ the mass flow rate of that liquid through the pipe in kg/s. The law for the conservation of momentum is

$$\delta PA = \frac{d(Mw)}{dt} = M \frac{dw}{dt} = \frac{M}{A\rho} \frac{dm}{dt} = L \frac{dm}{dt} \tag{2.2-1.21}$$

$$\delta P = \frac{L}{A} \frac{dm}{dt} \quad . \tag{2.2-1.22}$$

The upper equation follows directly from (2.2-1.4) if we neglect pressure drops and introduce the symbol $(\delta P = P_1 - P_2)$ for the pressure difference that causes flow change with gradient dm/dt. The coefficient beside dm/dt is the measure of that change, and it is called inertance I. The larger I is, the smaller the increase in mass flow for a given pressure difference δP.

Inertance I is thus defined as the ratio of effort (potential) δP to the gradient of change in fluid flow dm/dt.

$$I = \frac{\delta P}{\dfrac{dm}{dt}} = \frac{L}{A} \quad . \tag{2.2-1.23}$$

It is not difficult to see that the ratio I/R has the dimension of time, and thus we introduced in to Equation (2.2-1.20) the symbol for constant T

$$T = \frac{I}{R} = \frac{L}{AR} = \frac{L\Delta m}{AR\Delta m} = \frac{LA\rho\Delta w}{A\Delta\delta P} = \frac{M\Delta w}{A\Delta\delta P} \qquad (2.2\text{-}1.24)$$

If we expand the numerator and the denominator in the last equation by Δm, we get the meaning of the time constant

$$T = \frac{M\Delta w}{A\Delta\delta P} = \frac{\text{change in momentum}}{\text{change in force}} \qquad (2.2\text{-}1.25)$$

If dependence between pressure drop and mass flow was linear, and not quadratic as in the case of valves and pipes, we would have the following expression for resistance

$$R = \frac{\delta P}{m} \quad . \qquad (2.2\text{-}1.26)$$

Then we could show that

$$T = \frac{I}{R} = \frac{I}{R_v + R_c + |R_p|} = \frac{L}{A(R_v + R_c + |R_p|)}\,\frac{\overline{m}}{\overline{m}} = \frac{AL\rho\ \overline{w}}{A(\delta\overline{P}_v + \delta\overline{P}_c + |\delta\overline{P}_p|)} =$$
$$= \frac{M\overline{w}}{A\delta\overline{P}} = \frac{\text{momentum}}{\text{force}} \qquad (2.2\text{-}1.27)$$

This ends the analysis of liquid flow dynamics and we will not attempt to give the state-space equations. That will be done at the end of Section 2.2 for the most general case of an arbitrary fluid flow, for which this flow in Example 1 is a special case with the assumption $C = 0$. It is, however, useful to give a block diagram of variable interrelations obtained after the Laplace transformation of Equation (2.2-1.19). This has been done on Figure 2.2-1.4.

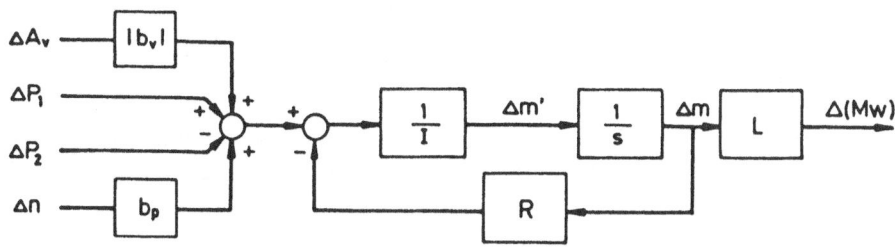

Fig. 2.2-1.4 Block diagram of the dynamics of liquid flow

The following two numerical examples will show the values of time constants characterizing liquid flow processes. We will show that these are fast processes with a small T, whose dynamics can usually and most frequently be neglected in comparison with the dynamics of the much slower final control elements (actuators).

Before analyzing numerical examples we must repeat that in the case of dynamic processes the most sensitive part is, as a rule, to determine the coefficients (parameters) of the dynamic model using statical (design) calculations. Here this concerns determining the coefficients R_p, R_v, R_c, b_p and b_v. To calculate them correctly it is necessary to know very well the methods of statical calculation, as the following lines will show.

Example 2 Calculation of dynamic coefficients. R, I, T

Cold water (ρ = 1000 kg/m^3, η = 0.001 kg/ms) runs at a rate of 0.417 kg/s (1.5 m^3/h) through a pipe of inner diameter 25 mm and length

a) 100 m
b) 1000 m.

The pipe is smooth, with two elbows and control valve NO 25 in which the pressure drop for the given flow is 0.5 bar. The flow is not forced by pump but is maintained by pressure difference between the tanks. Determine time constants T.

a) L = 100 m, A = $d^2\pi/4$ = 0.0005 m^2

$$T = \frac{L}{A(R_v+R_c+|R_p|)} \qquad\qquad (2.2\text{-}1.28)$$

The above expression shows that it is still necessary to determine the resistances. As there is no pump, R_p = 0. From (2.2-1.13)

$$R_v = \frac{2\delta\bar{P}_v}{m} = 2.3981 \ \frac{bar}{kg/s} \ .$$

For the pipe and the built-in elbows there is the already known formula for pressure drop calculation

$$\delta P_c = (\lambda \frac{L}{d} + \xi) \frac{\rho w^2}{2} = \underbrace{\frac{\lambda \frac{L}{d} + \xi}{2 A^2 \rho}}_{K_c} \; m^2 \qquad\qquad (2.2\text{-}1.29)$$

K_c is the coefficient used in Equation (2.2-1.8). The unknown in the expression for δP_c is the coefficient λ, which depends only on the Reynolds number for a smooth pipe and is calculated from the Blasius formula

$$Re = \frac{wd\rho}{\eta} = \frac{0.85 \cdot 0.025 \cdot 100}{0.001} = 21250 \; > \; 2320$$

$$\lambda = \frac{0.3164}{\sqrt[4]{Re}} = 0.0261 \quad .$$

The individual resistances coefficient for the elbow is 0.2, and can be written as follows

$$K_c = \frac{0.0261 \dfrac{100}{0.025} + 2 \cdot 0.2}{2 \cdot 1000 \cdot 0.0005^2} = 217466 \quad .$$

$$\delta P_c = 217466 \; \overline{m}^2 = 37815 \; \frac{N}{m^2} \quad .$$

Referring to (2.2-1.15) we have

$$R_c = \frac{2\delta P_c}{\overline{m}} = \frac{2 \cdot 37815}{0.417} = 1.814 \; \frac{bar}{kg/s}$$

Now we can get from (2.2-1.28)

$$T = \frac{L}{A(R_v + R_c)} = \frac{100 \cdot 10^{-5}}{0.0005(2.398 + 1.814)} = 0.474 \; s \quad .$$

b) $L = 1000$ m, $A = 0.0005$ m^2

The values for R_p and R_v do not change, and R_c and T are calculated as follows

$$K_c = \frac{0.0261 \dfrac{1000}{0.025} + 2 \cdot 0.2}{2 \cdot 0.0005^2 \cdot 1000} = 20.9 \cdot 10^5 \quad .$$

$$\delta P_c = 363219 \; \frac{N}{m^2} \quad . \quad R_c = \frac{2\delta P_c}{\overline{m}} = 17.42 \; \frac{bar}{kg/s} \quad .$$

$$T = \frac{1000}{0.0005(2.3981+17.42)10^5} = 1.009 \ s \quad .$$

Therefore, although the length of the pipe increased 10 times, the time constant T is only twice bigger. The reason for this is that as length L increases (increased inertia) so does also the resistance of the pipe R_c, which decreases the growth of the time constant T.

Example 3 Calculation of dynamic coefficients. R, I, T

Water flows at 3 m/s through a pipe 300 mm in diameter and 300 m long. The relative roughness of the pipe is $\epsilon/d = 0.002$, and the kinematic viscosity of water $\nu = 9 \cdot 10^{-7} \ m^2/s$, $\rho = 1000 \ kg/m^3$. Calculate the time constant T.

As there is no pump and valve in the pipe

$$T = \frac{I}{R_c} = \frac{L}{AR_c} \quad .$$

To determine the pipe resistance R_c we must first calculate the pressure drop along the pipe. $A = 0.0707 \ m^2$, $m = Aw\rho = 212 \ kg/s$, $Re = wd/\nu = 10^6$, the coefficient λ for rough pipes can be determined either from the Altshul or the Moody formula

Moody: $\lambda = 0.0055 \left[1+(20000 \frac{\epsilon}{d} + \frac{10^6}{Re})^{\frac{1}{3}} \right] = 0.0245$

Altshul: $\lambda = 0.11 \ (\frac{\epsilon}{d} + \frac{68}{Re})^{0.25} = 0.02346$

For this field of Re and relative roughness the difference is only 4%, and in the further calculations we will use the value $\lambda = 0.02346$.

$$K_c = \lambda \frac{L}{d} \frac{1}{2A^2\rho} = 2.346 \quad ,$$

$$\delta P_c = K_c \overline{m}^2 = 2.346 \cdot 212^2 = 105438 \ Pa \quad ,$$

$$R_c = \frac{2\delta P_c}{m} = \frac{2 \cdot 105438}{212} = 994.7 \ \frac{Pa}{kg/s} \quad ,$$

$$T = \frac{300}{0.0707 \cdot 994.7} = 4.265 \ s \quad .$$

2.2-2 GAS AND STEAM FLOW

As in the case with liquids, here we will also begin by making as many assumptions as possible to obtain a simple dynamic model for the flow rate and pressure in gas pipes. If we prune the physical process, in this way, of phenomena that are at present secondary, insight into its basic dynamic properties will be easier.

The analysis that follows is true of relatively short pipes through which gas (steam) flows with the usual technical velocities of flow (much smaller than critical), and along which pressure drop resulting from various resistances never exceeds 10% of the pressure. In this case pressure drop can always be calculated as in the case of liquid. These are the introductory assumptions that will be valid for the complete following analysis.

Unlike in the case of liquids, changes in gas density ρ **can now not be neglected,** but because of low gas (steam) density **the inertia of the gas (steam) mass in the pipe can be neglected.** Therefore, we leave inertia effects out of our analysis $(I = 0)$, and take into consideration possibilities of mass storage in the pipe because of the compressibility of the media $(\rho \neq \text{const.}, C \neq \text{const.})$. As in the case of liquids, gas flow also has resistances along the pipe: valves, flanges and the like. From the aspect of gas state changes in the pipe, we will assume here the case of **isothermal** change. There is no special difficulty if we want to consider polytropic change. In that case we must simply insert in all the equations, instead of the isothermal capacitance C, the polytropic capacitance C_n. As in the case of the tank, here also we will use the state equations for ideal gas.

Example 1 Dynamics of gas or steam flow. Boundary conditions

A compressor (air pump) drives m kg/s of gas through a control valve between two tanks. Derive the mathematical model describing the dynamic changes in mass flow and pressure in the pipe.

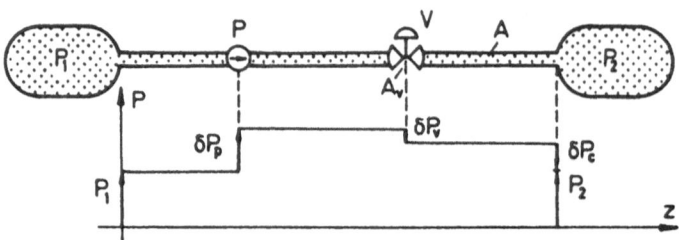

Fig. 2.2-2.1 Gas (steam) transport between two tanks

The pipe has two properties. Together with all the fittings and the pump, it resists gas flow. This resistance is characterized by total resistance R. It also enables storage of a mass of gas, i.e. it has the property of capacitance C. Thus this pipe can be represented as a connection in series of resistance and capacity, which has been done on the following two figures.

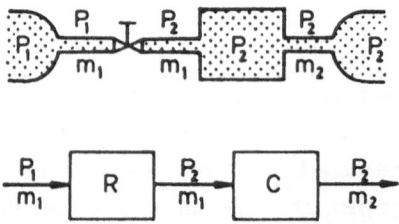

Fig. 2.2-2.2 Pipe approximation in R-C series

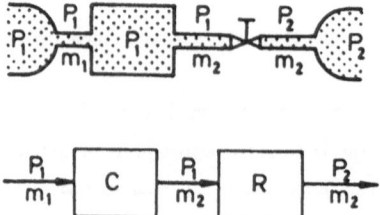

Fig. 2.2-2.3 Pipe approximation in C-R series

In the upper figures, R represents all the possible resistances that influence pressure change on the observed length of pipe L, and C represents the total capacitance of the whole system.

Although this does not seem important at first glance, we must nevertheless emphasize the following. The order in the series **is not arbitrary** but depends on the given boundary conditions at the ends of the pipe. That this is so will be shown in the following lines when we formulate equations for both the arrangements. As the possibilities of mass storage are not neglected, we will get the desired dynamic models by formulating **equations for the conservation of mass** and **momentum,** and in the latter we will neglect the momentum of the mass of gas in the pipe.

For the R-C order on Figure 2.2-2.2 we have

$$m_1 - m_2 = \frac{dM}{dt} = \frac{V}{R\vartheta} \frac{dP_2}{dt} \quad . \tag{2.2-2.1}$$

$$P_1 + \delta P_p - \delta P_v - \delta P_c - P_2 = \frac{d(Mw)}{dt} = 0 \quad . \tag{2.2-2.2}$$

$$P_1 - P_2 = \delta P_v + \delta P_c - \delta P_p \quad . \tag{2.2-2.3}$$

After we neglect the influence of changes in the number of revolutions of the compressor Δn and changes in the cross-sectional area of the valve ΔA_v (which does not make us lose anything essential but only makes it easier to derive the equations), the linearization of the upper equations gives

$$\Delta m_1 - \Delta m_2 = C \frac{d\Delta P_2}{dt} \quad , \tag{2.2-2.4}$$

$$\Delta P_1 - \Delta P_2 = (R_v + R_c + |R_p|) \Delta m_1 = R\Delta m_1 \tag{2.2-2.5}$$

The last equation, except for the assumptions made here ($\Delta n = 0$, $\Delta A_v = 0$, $I = L/A = 0$), is identical to Equation (2.2-1.19).

For the C-R order we analogously get from Figure 2.2-2.3

$$\Delta m_1 - \Delta m_2 = C\frac{d\Delta P_1}{dt} \tag{2.2-2.6}$$

$$\Delta P_1 - \Delta P_2 = R \Delta m_2 \quad . \hspace{6cm} (2.2\text{-}2.7)$$

The symbols C and R have the same meaning as in the preceding chapters

$$C = \frac{V}{R \vartheta} = \frac{\overline{M}}{\overline{P}} \quad , \quad R = R_v + R_c + |R_p| = \frac{2 \delta \overline{P}_v}{m} + \frac{2 \delta \overline{P}_c}{m} + |R_p|$$

Equations (2.2-2.4) - (2.2-2.7) are pairs of interconnected equations with four variables - Δm_1, Δm_2, ΔP_1 and ΔP_2. To solve those pairs of equations it is necessary to know two variables, and then it is no problem to determine the other two. For fluid flow processes we can make an arbitrary selection of independent boundary conditions (inputs), except that **mass flow rate and pressure** (both boundary conditions) **must not be given at the same end, at the same time.** (We will say more about boundary conditions in Chapter 3, when we talk about partial differential equations. For the present it is enough to say that if both the boundary conditions are given at the same end, the other end remains completely undetermined because there is no information about the way in which that end of the pipe "contacts" the surroundings). Figure 2.2-2.4 shows the four possible cases of boundary conditions that can appear in practice. The most frequent and most interesting are examples a) and b), and here we will analyze these cases in more detail.

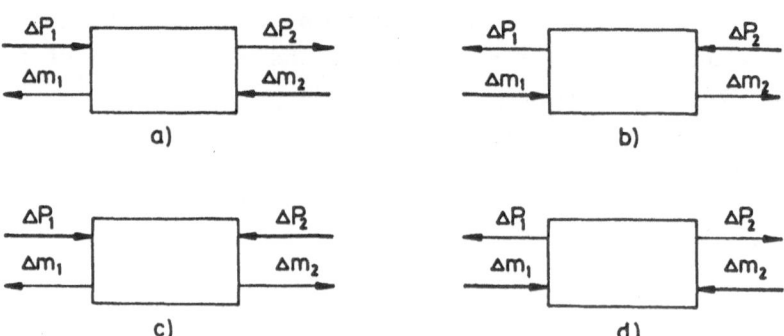

Fig. 2.2-2.4 Possible boundary conditions in fluid flow

The following derivations will show the truth of the statement that in a dynamic representation the order of resistance and capacity elements is determined by the given boundary conditions.

a) ΔP_1 and Δm_2 are given. Determine ΔP_2 and Δm_1.

- R-C order. Figure 2.2-2.2 .

Equations (2.2-2.4) and (2.2-2.5) give

$$T \frac{d\Delta m_1}{dt} + \Delta m_1 = \Delta m_2 + C \frac{d\Delta P_1}{dt} \; . \qquad\qquad (2.2\text{-}2.8)$$

$$T \frac{d\Delta P_2}{dt} + \Delta P_2 = - R\Delta m_2 + \Delta P_1 \; . \qquad\qquad (2.2\text{-}2.9)$$

- C-R order. Figure 2.2-2.3 .

Equations (2.2-2.6) and (2.2-2.7) give

$$\Delta m_1 = \Delta m_2 + C \frac{d\Delta P_1}{dt} \; . \qquad\qquad (2.2\text{-}2.10)$$

$$\Delta P_2 = -R\Delta m_2 + \Delta P_1 \; . \qquad\qquad (2.2\text{-}2.11)$$

$$T = CR$$

The last four equations show that the boundary conditions on Figure 2.2-2.4a, depending on the order of resistance and capacity elements, give completely different models for pressure and flow dynamics. The C-R order is obviously not satisfactory because in it the components of time lag, $Td\Delta m_1/dt$ and $Td\Delta P_2/dt$ are lost. For the second pair of boundary conditions we can draw similar conclusions.

b) Δm_1 and ΔP_2 are given. Determine Δm_2 and ΔP_1.

- R-C order. Figure 2.2-2.2 .

$$\Delta m_2 = \Delta m_1 - C \frac{d\Delta P_2}{dt} \; . \qquad\qquad (2.2\text{-}2.12)$$

$$\Delta P_1 = R\Delta m_1 + \Delta P_2 \; . \qquad\qquad (2.2\text{-}2.13)$$

- C-R order. Figure 2.2-2.3 .

$$T \frac{d\Delta m_2}{dt} + \Delta m_2 = \Delta m_1 - C \frac{d\Delta P_2}{dt} \; . \qquad\qquad (2.2\text{-}2.14)$$

$$T \frac{d\Delta P_1}{dt} + \Delta P_1 = R\Delta m_1 + \Delta P_2 \qquad\qquad (2.2\text{-}2.15)$$

Now the R-C order does not reproduce the dynamics faithfully enough, because time lag elements have disappeared from its dynamic model.

Therefore, when we approximate a pipe with elements of resistance and capacity, we must take care of the boundary conditions. Where a pressure change is given we must put an element of resistance, and on the end where flow rate is given we put capacity.

In practice we often use the solution that the total pipe resistance is divided into input and output pipe resistance with a capacity element in between. This arrangement is shown on Figure 2.2-2.5.

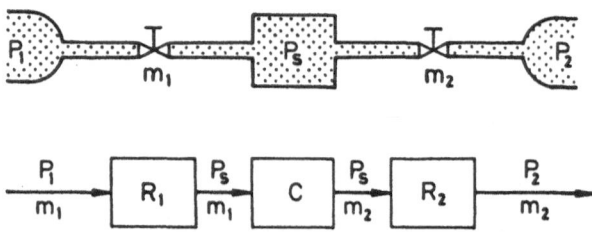

Fig. 2.2-2.5 Pipe approximated with input and output resistance and capacity element

The equation for the conservation of mass and two equations for the conservation of momentum for the order shown on Figure 2.2-2.5, in linear form, are as follows

$$\Delta m_1 - \Delta m_2 = C \frac{d\Delta P_s}{dt} \quad , \tag{2.2-2.16}$$

$$\Delta P_1 - \Delta P_s = R_1 \Delta m_1 \quad , \tag{2.2-2.17}$$

$$\Delta P_s - \Delta P_2 = R_2 \Delta m_2 \tag{2.2-2.18}$$

c) Δm_1 and ΔP_2 are given. Determine Δm_2 and ΔP_1.

- R_1 - C - R_2 order, Figure 2.2-2.5 .

Rearranging the last three equations we get

$$T_2 \frac{d\Delta m_2}{dt} + \Delta m_2 = \Delta m_1 - C \frac{d\Delta P_2}{dt} \ , \tag{2.2-2.19}$$

$$T_2 \frac{d\Delta P_1}{dt} + \Delta P_1 = R_1 T_2 \frac{d\Delta m_1}{dt} + (R_1 + R_2)\Delta m_1 + \Delta P_2 \tag{2.2-2.20}$$

$$T_2 = R_2 C$$

If we compare the last two equations with Equations (2.2-2.14) and (2.2-2.15), we see the following differences. Time constant T_2 is smaller than T, and there is also a derivative dependence of pressure P_1 on mass flow rate m_1. We thus have a property that is lost in the R-C combination. Resistance division is already a step towards distribution, towards something that is in fact distributed along the pipe, but in this part of the book lumped in one place. Thus we must believe that the approximation in Figure 2.2-2.5 is better than that in Figure 2.2-2.2, and the model in the form of the last two equations closer to the real behavior of flow rate Δm_2 and pressure ΔP_1.

d) ΔP_1 and Δm_2 are given. Determine ΔP_2 and Δm_1.

- order R_1 - C - R_2, Figure 2.2-2.5 .

After rearrangement Equations (2.2-2.16) - (2.2-2.18) get the desired form

$$T_1 \frac{d\Delta m_1}{dt} + \Delta m_1 = \Delta m_2 + C \frac{d\Delta P_1}{dt} \ . \tag{2.2-2.21}$$

$$T_1 \frac{d\Delta P_2}{dt} + \Delta P_2 = \Delta P_1 - (R_2 T_1 \frac{d\Delta m_2}{dt} + (R_1 + R_2)\Delta m_2) \tag{2.2-2.22}$$

$$T_1 = R_1 C$$

Now T_1 is smaller than T in Equations (2.2-2.8) and (2.2-2.9), and like in c) there is a derivative dependence of pressure P_2 on mass flow rate change m_2.

Here we must be completely clear and express our feeling that approximation will be better if we divide the pipe into even more resistance and capacity elements, but the addition of new capacity elements would increase the order of the system. We must also say that there are no exact analytical criteria about when and into how many elements specific pipes should be divided. In these contemplations we can use the fact that time constants linked with gas and steam flow are small and in practice rougher approximations (R-C or C-R), or even statical relations, will satisfy. What has just been said is true unless we have periodic (oscillatory) processes, which are not included in this model because we neglected inertia (I = 0).

Finally, it is also useful to represent the dynamic relations between input and output variables in the model given by Equations (2.2-2.21) and (2.2-2.22) by matrix transfer functions.

$$
\begin{bmatrix} \Delta P_2(s) \\ \\ \Delta M_1(s) \end{bmatrix} = \begin{bmatrix} \dfrac{1}{T_1 s+1} & -R\,\dfrac{Ts+1}{T_1 s+1} \\ \\ \dfrac{Cs}{T_1 s+1} & \dfrac{1}{T_1 s+1} \end{bmatrix} \begin{bmatrix} \Delta P_1(s) \\ \\ \Delta M_2(s) \end{bmatrix} \qquad (2.2\text{-}2.23)
$$

$$\mathbf{Y}(s) = \mathbf{G}(s)\ \mathbf{U}(s)$$

$$R = R_1 + R_2 \quad,\quad T = \frac{R_2}{R}\, T_1$$

The upper expression, if we put $R_1 = R$ and $R_2 = 0$, from which follows $T = 0$, is a matrix presentation of case a) in this example.

⎯⎯⎯⎯⎯⎯⎯⎯

Before we determine the numerical value of the time constant for an air pipe in the following example, we must say that there is a complete analogy between the dynamics of a gas tank and a pipe through which gas flows if there are very small pressure differences in the pipe and if the flow is far from critical. With regard to the value of the time constant, it will be smaller than the constant that characterized the gas or steam tank dynamics, not only because of the usually smaller capacity but also because of the much smaller resistance the pipe offers to flow.

⎯⎯⎯⎯⎯⎯⎯⎯

Example 2 Calculation of dynamic coefficients. R, C, T

Air of density ρ = 1.16 kg/m^3 and kinematic viscosity ν = 15.7·10^{-6} m^2/s flows at a velocity of 20 m/s through a pipe of diameter 200 mm and length 100 m. The pressure in the pipe equals external pressure and temperature $\vartheta = 0$ °C; the pipe roughness is $\epsilon = 0.1$ mm. Determine time constant T that characterizes the process of flow and pressure change in the pipe.

As the pipe has no pump or valve, T will be determined only by resistance to flow through the pipe, $T = R_c C$.

For flow through the pipe and for small pressure differences the calculation of the pressure drop due to friction in the pipe runs in the same way as in the case of liquid. The calculation itself follows, in which λ is calculated from the Altshul formula.

$$\delta \overline{P}_c = \lambda \frac{L}{d} \frac{\overline{m}^2}{2A^2\overline{\rho}} = K_c \overline{m}^2 \quad ,$$

$$R_c = \frac{2 \delta \overline{P}_c}{m} \quad ,$$

$$Re = \frac{wd}{\nu} = \frac{20 \cdot 0.2}{15.7 \cdot 10^{-6}} = 0.255 \ 10^6 \quad ,$$

$$A = \frac{d^2 \pi}{4} = 0.0314 \ m^2$$

$$m = A w \rho = 0.7285 \ \frac{kg}{s} \quad ,$$

$$\lambda = 0.11 \ (\frac{\epsilon}{d} + \frac{68}{Re})^{0.25} = 0.0174 \quad ,$$

$$\delta P_c = 0.0174 \ \frac{100}{0.2} \ \frac{0.7285^2}{2 \cdot 0.00314^2 \cdot 1.16} = 2015.68 \ \frac{N}{m^2} \quad ,$$

$$R_c = 5533.8 \ \frac{N/m^2}{kg/s} \quad ,$$

$$C = \frac{V}{R\vartheta} = \frac{3.14}{287 \cdot 292} = 3.747 \cdot 10^{-5} \frac{kg \ m^2}{N} \quad ,$$

$$T = R_c \ C = 0.2073 \ s \quad .$$

2.2-3 FLUID FLOW

In the previous analysis of liquid, gas and steam dynamics the models were obtained after assumptions that narrowed down their field of application and values. However, those models made sense and can

answer questions concerning basic dynamic processes for very many
devices and plants. Liquids, however, do possess the property of
compressibility, which was neglected in 2.2-1 by making C = 0. The
consequences of that property are the known phenomena of periodic
oscillations of the liquid column, i.e. water hammer effects. Gases, in spite
of their small density ρ, nevertheless possess kinetic energy, and
momentum (Mw) during their flow, which was made equal to zero in 2.2-2,
cannot always be neglected. In short, it is desirable to derive a model for
dynamic changes of mass flow rate and pressure in a pipe for an
arbitrary fluid, not neglecting either its compressibility or its
momentum, i.e. the kinetic energy a flowing mass of fluid possesses. This
will be done in the following lines, and it will be shown that the
preceding models are only special cases of the one obtained for an
arbitrary fluid with properties of capacitance (C), resistance (R) and
inertance (I).

Also, when the processes of fluid flow are later regarded as processes
with distributed parameters in Section 3.3, we will get models whose first
and roughest approximation will be equal (or, at least, similar) to the
equations derived here, for which all the properties were considered
lumped in one point in space.

Example 1 Dynamics of fluid flow. Periodic processes

Through a pipe with solid walls, of length L and cross-sectional area
A, flows m kg/s fluid at a velocity w, density ρ and temperature ϑ.

1) Derive the linear model for mass flow rate and pressure changes at
the inlet and outlet cross-section for the following boundary conditions:

a) ΔP_1 and Δm_2 are given.
b) ΔP_2 and Δm_1 are given.

2) For case a) derive a model in the form of matrix state-space
equations, and a matrix transfer function representation.

1a) For the given input variables the pipe can be shown by an
arrangement of elements as presented on Figure 2.2-3.1.

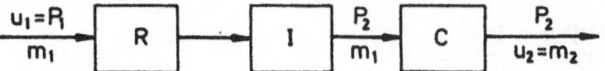

Fig. 2.2-3.1 Fluid flow with the properties of resistance,
inertance and capacitance

Here also, analogously to the earlier procedures, we formulate
equations for the conservation of mass and momentum

$$m_1 - m_2 = \frac{dM}{dt} = \frac{d(V\rho_2)}{dt} = \frac{V}{R\vartheta}\frac{dP_2}{dt} \qquad (2.2\text{-}3.1)$$

$$(P_1 + \delta P_p - \delta P_v - \delta P_c - P_2) A = \frac{d(Mw_1)}{dt} =$$

$$= AL\rho\frac{dw_1}{dt} = L\frac{dm_1}{dt} \quad . \qquad (2.2\text{-}3.2)$$

The upper equations are nonlinear and if they are linearized (again
taking $\Delta n = 0$, $\Delta A_v = 0$), we get the following two equations

$$\Delta m_1 - \Delta m_2 = C\frac{d\Delta P_2}{dt} \quad , \qquad (2.2\text{-}3.3)$$

$$\Delta P_1 - \Delta P_2 = R\Delta m_1 + \frac{L}{A}\frac{d\Delta m_1}{dt} \quad . \qquad (2.2\text{-}3.4)$$

$$I = \frac{L}{A}$$

In the general case $R = R_c + R_v + |Rp|$, compare the derivations for
liquid flow from Section 2.2-1. The upper two equations give the desired
model in the following form

$$IC\frac{d^2\Delta P_2}{dt^2} + RC\frac{d\Delta P_2}{dt} + \Delta P_2 = \Delta P_1 - (I\frac{d\Delta m_2}{dt} + R\Delta m_2) \quad . \qquad (2.2\text{-}3.5)$$

$$IC\frac{d^2\Delta m_1}{dt^2} + RC\frac{d\Delta m_1}{dt} + \Delta m_1 = C\frac{d\Delta P_1}{dt} + \Delta m_2 \quad . \qquad (2.2\text{-}3.6)$$

For the first time in this book we have obtained **differential
equations of the second order.** There are thus **two** possible **mass
or energy storage elements.** Truly, fluid is compressible and fluid
mass can be stored in the given volume V. This property is represented

by element C on Figure 2.2-3.1. But fluid mass storage also represents potential energy storage, as will be seen from further discussion. Equation (2.2-3.1) can also be written

$$m_1 - m_2 = \frac{d(VP_2)}{dt} \frac{1}{R\vartheta} \quad . \tag{2.2-3.7}$$

The term VP_2 has the dimension of energy Nm and is a measure for potential energy change in the pipe volume V resulting from pressure change P_2.

Similarly, an analysis of the equation for the conservation of momentum indicates changes in kinetic energy. If we introduce new symbols, (2.2-3.2) can also be written

$$\delta PA = F = M \frac{dw_1}{dt} \quad . \tag{2.2-3.8}$$

δP denotes the total pressures difference acting on the fluid mass in the pipe, and F the force that causes the displacement of that fluid. If the fluid flows at the velocity w_1, then it passes the path $dz = w_1 dt$ in a time period dt. If the last equation is multiplied by dz, we get

$$Fdz = M \frac{dw_1}{dt} w_1 dt = M(dw_1) w_1 = \frac{1}{2} M d(w_1^2) \quad . \tag{2.2-3.9}$$

Let the fluid mass, which is in position z_0 at the moment t_0, move into position z under the influence of force F, and let its velocity changes from w_{10} to w_1. Integration of the upper equation gives

$$\int_{z_0}^{z} Fdz = \frac{M}{2} \int_{w_{10}}^{w_1} d(w_1^2) \quad , \tag{2.2-3.10}$$

$$W = \frac{Mw_1^2}{2} - \frac{Mw_{10}^2}{2} = \Delta E_k \quad . \tag{2.2-3.11}$$

The integral on the left-hand side is work W that the pressure forces would realize if mass M moved from z_0 to z, and that work would be transformed into kinetic energy. In the case of a pipe of length L, changes in any of the pressures (which represent a change in force F that acts on the fluid) do not cause the mass of fluid M to move, but lead to the compression or expansion of the fluid in pipe volume V. This, in turn, leads to an energy-form change-kinetic into potential for compression and potential into kinetic for expansion.

It is not difficult to see that this is an oscillatory (periodic) process. Let m_2 increase by $\Delta m_2 = 1$ at $t = 0$. From (2.2-3.7) pressure gradient P_2 becomes negative and P_2 (the potential energy of the fluid in the pipe) decreases. This makes the total pressure difference δP increase, and from (2.2-3.8) the velocity w_1 of the fluid at the entrance increases (i.e. its kinetic energy increases). Because of an increase in m_1 gradient $d\Delta P_2/dt$ remains negative, but its absolute value decreases and equals zero at the moment when $m_1 = \overline{m} + \Delta m_2$, i.e. for $\Delta m_1 = \Delta m_2$. The process does not end here because pressure P_2 has decreased by ΔP_2 and δP is still positive: consequently the velocity w_1 (the kinetic energy of the fluid) continues to grow. Kinetic energy will increase until P_2 returns to its initial steady value. At that moment $\delta P = \delta \overline{P}$, or $\Delta \delta P = 0$ and $d(\Delta w_1)/dt = 0$. However, the process does not end here either, because now m_1 kg/s of fluid flows in from the outside with maximum velocity $w_1 = \overline{w}_1 + \Delta w_1$. Since $m_1 > m_2$, according to (2.2-3.7) we have an increase in P_2. This process continues to repeat itself until friction reduces energy changes to zero and Δm_1 becomes equal to Δm_2.

Figure 2.2-3.2 gives a qualitative presentation of changes in the variables Δm_1 and ΔP_2 in real conditions (friction present) for the described case of increase in fluid consumption in the outlet section for $\Delta m_2 = 1$ kg/s. Figure 2.2-3.3 gives the course of the same variables in the case of pressure increase $\Delta P_1 = 1$ bar at the pipe inlet.

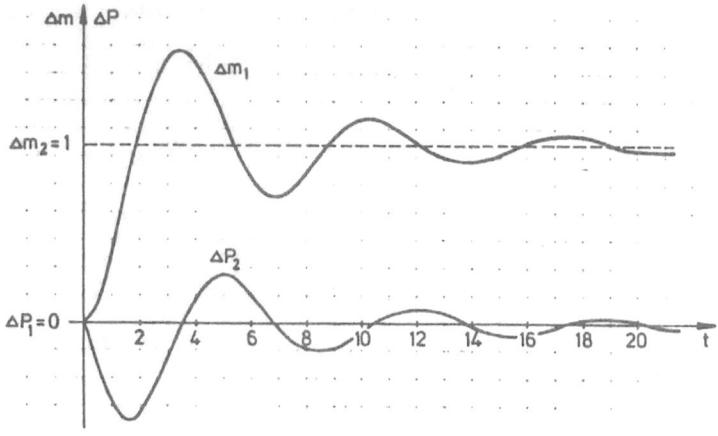

Fig. 2.2-3.2 Periodic changes in mas flow rate at pipe inlet Δm_1 and pressure at pipe outlet ΔP_2 in the case of increased fluid consumption $\Delta m_2 = 1$ kg/s

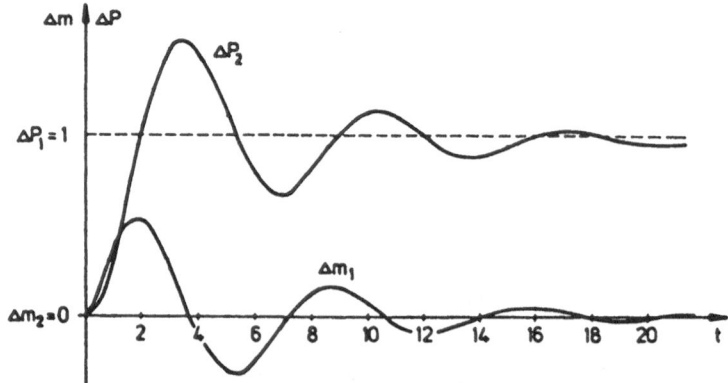

Fig. 2.2-3.3 Periodic changes in mass flow rate at pipe inlet Δm_1 and
pressure at pipe outlet ΔP_2 in the case of increased inlet
pressure ΔP_1

The following shorter derivation shows that if there is **no friction,** i.e.
no energy loss, state changes of fluid in a pipe really conserve the
energy (it only changes in form) the fluid contains within it. With R = 0,
Equations (2.2-3.3) and (2.2-3.4) become

$$\Delta m_1 - \Delta m_2 = C \frac{d\Delta P_2}{dt} \quad . \tag{2.2-3.12}$$

$$\Delta P_1 - \Delta P_2 = I \frac{d\Delta m_1}{dt} \quad . \tag{2.2-3.13}$$

If we divide the upper equations one by the other we get

$$\frac{\Delta m_1 - \Delta m_2}{\Delta P_1 - \Delta P_2} = \frac{C}{I} \frac{d\Delta P_2}{d\Delta m_1} \tag{2.2-3.14}$$

If outside disturbances do not influence fluid in the pipe, i.e. with Δm_2
= 0 and ΔP_1 = 0, (2.2-3.14) can also be written

$$\frac{I}{2\rho} d(\Delta m_1{}^2) + \frac{C}{2\rho} d(\Delta P_2{}^2) = 0 \quad . \tag{2.2-3.15}$$

In the last equation we divide both sides by fluid density $\bar{\rho}$ in the
steady state. This, of course, does not change the result, but leads to the
desired form for the final expression. From that equation it follows directly
that

$$\underbrace{\frac{I_\rho \Delta m_1^2}{2}}_{KE} + \underbrace{\frac{C_\rho \Delta P_2^2}{2}}_{PE} = \text{const} = E_0 \quad . \qquad\qquad (2.2\text{-}3.16)$$

$$\left(I_\rho = \frac{I}{\rho} \ , \ C_\rho = \frac{C}{\rho} \right)$$

The upper equation is the law for the conservation of energy. A dimensional analysis shows that both the terms on the left-hand side have the dimension of energy (Nm), the first term representing kinetic energy change, and the second potential energy change in the fluid. In this frictionless case the periodic responses shown in Figures 2.2-3.2 and 2.2-3.3 would not be damped and oscillations would continue, since the initial energy E_0 would always be preserved in the fluid, with undamped amplitudes until infinity.

Equations (2.2-3.5) and (2.2-3.6) are the most general form of the model for fluid flow dynamics from which models for gas or steam flow (I = 0) or for fluid flow (C = 0) are obtained directly. If we insert I = 0 into those two equations we get Equations (2.2-2.8) and (2.2-2.9) from the section on gas flow. If we want to observe changes in liquid flow due to inlet and outlet pipe pressure changes, inserting C = 0 into (2.2-3.5) gives us a model like in Equation (2.2-1.19).

1b) If Δm_1 and ΔP_2 are given, the following sequence of elements makes it easier to get a dynamic model for fluid state changes.

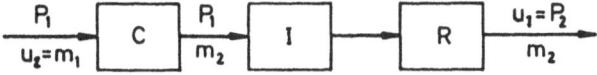

Fig. 2.2-3.4 Fluid flow with the properties of capacitance, inertance and resistance

Using the same assumptions as in case a), and analogously with Equations (2.2-3.3) and (2.2-3.4), we can write

$$\Delta m_1 - \Delta m_2 = C \frac{d \Delta P_1}{dt} \quad , \tag{2.2-3.17}$$

$$\Delta P_1 - \Delta P_2 = R \Delta m_2 + \frac{L}{A} \frac{d \Delta m_2}{dt} \quad . \tag{2.2-3.18}$$

Rearrangement of those equations gives the desired expressions

$$IC \frac{d^2 \Delta P_1}{dt^2} + RC \frac{d \Delta P_1}{dt} + \Delta P_1 = \Delta P_2 + I \frac{d \Delta m_1}{dt} + R \Delta m_1 \quad , \tag{2.2-3.19}$$

$$IC \frac{d^2 \Delta m_2}{dt^2} + RC \frac{d \Delta m_2}{dt} + \Delta m_2 = \Delta m_1 - C \frac{d \Delta P_2}{dt} \quad . \tag{2.2-3.20}$$

2) To represent the model from 1a) in the form of state-space equations we must first determine which variables are considered inputs and which outputs, and which represent state variables. Now this selection is not difficult as it is completely natural for the input vector to be composed of ΔP_1 and Δm_2, and the state and output vector of ΔP_2 and Δm_1. We will thus have

$$\mathbf{X} = \mathbf{Y} = \begin{bmatrix} \Delta P_2 \\ \Delta m_1 \end{bmatrix} \quad , \qquad \mathbf{U} = \begin{bmatrix} \Delta P_1 \\ \Delta m_2 \end{bmatrix} \quad .$$

In formulating matrices $\mathbf{A}$, $\mathbf{B}$ and $\mathbf{C}$ ($\mathbf{D} = 0$ because there is no direct action of $\mathbf{U}$ on $\mathbf{Y}$), it is best to start from Equations (2.2-3.3) and (2.2-3.4) from which we immediately get expressions for $\Delta P_2'$ and $\Delta m_1'$. Written differently, those equations are

$$\begin{bmatrix} \Delta P_2 \\ \Delta m_1 \end{bmatrix}' = \begin{bmatrix} 0 & \frac{1}{C} \\ -\frac{1}{I} & -\frac{R}{I} \end{bmatrix} \begin{bmatrix} \Delta P_2 \\ \Delta m_1 \end{bmatrix} + \begin{bmatrix} 0 & -\frac{1}{C} \\ \frac{1}{I} & 0 \end{bmatrix} \begin{bmatrix} \Delta P_1 \\ \Delta m_2 \end{bmatrix} \quad . \tag{2.2-3.21}$$

$$\mathbf{X'} = \mathbf{A} \, \mathbf{X} + \mathbf{B} \, \mathbf{U} \quad .$$

$$\begin{bmatrix} \Delta P_2 \\ \Delta m_1 \end{bmatrix} = \begin{bmatrix} 1 & 0 \\ 0 & 1 \end{bmatrix} \begin{bmatrix} \Delta P_2 \\ \Delta m_1 \end{bmatrix} \quad . \tag{2.2-3.22}$$

$$\mathbf{Y} = \mathbf{C} \, \mathbf{X} \quad .$$

The eigenvalues of matrix $\mathbf{A}$ point to transient phenomena and it would be useful to determine expressions from which they can be calculated. From Equation $|\lambda \mathbf{I} - \mathbf{A}| = 0$ we get

$$\begin{bmatrix} \lambda & -\dfrac{1}{C} \\[2ex] \dfrac{1}{I} & \lambda + \dfrac{R}{I} \end{bmatrix} = \lambda^2 + \frac{R}{I}\lambda + \frac{1}{IC} = IC\,\lambda^2 + RC\lambda + 1 = 0 \qquad (2.2\text{-}3.23)$$

It should be observed that the upper equation is the same as the characteristic equation of the ODE (2.2-3.5) and (2.2-3.6).

The transient functions from Figures 2.2-3.2 and 2.2-3.3 and the differential equations (2.2-3.5) and (2.2-3.6) show that in this case we have the standard behavior of a second-order (proportional) system that shows periodic characteristics. We will, therefore, introduce symbols for natural oscillation frequency ω_n and the damping coefficient ξ.

$$\omega_n = \frac{1}{\sqrt{IC}} \quad . \qquad\qquad\qquad\qquad (2.2\text{-}3.24)$$

$$\xi = \frac{RC}{2\sqrt{IC}} = \frac{RC}{2}\,\omega_n \quad . \qquad\qquad\qquad (2.2\text{-}3.25)$$

With those variables the characteristic equation (2.2-3.23) can be written in the usual form for 2nd-order processes

$$\lambda^2 + 2\xi\omega_n\lambda + \omega_n^2 = 0 \quad . \qquad\qquad\qquad (2.2\text{-}3.26)$$

The eigenvalues are determined by the expression

$$\lambda_{1,2} = -\frac{R}{2I} \pm \sqrt{\frac{1}{IC}\left(\frac{R^2C^2}{4IC} - 1\right)} = -\xi\omega_n \pm \omega_n \sqrt{\xi^2 - 1} = \sigma \pm \omega j$$
$$(2.2\text{-}3.27)$$

Here we will not enter into a more detailed analysis of dynamic features because the key properties are completely obvious from the solution of the upper equation. For real cases λ will always have a negative real part, $\sigma < 0$, because $\xi\omega_n > 0$. In theoretic cases, when friction is neglected ($R = 0$), ξ equals zero and so does, therefore, the real part $\sigma = 0$, i.e. $\lambda_{1,2} = \omega_n j = \pm(IC)^{-0.5}$ j. The eigenvalues are conjugate-complex pairs on the imaginary axis and the system oscillates without damping with frequency ω_n. In the practically most frequent case we have $0 < \xi < 1$, and these are the already shown damped oscillations. Increasing the damping coefficient leads to a decrease in oscillation frequency, and in the case of $\xi \geq 1$ both mass flow and pressure changes are no longer periodical. The expression under the square root is no longer imaginary and the solution are two negative and real eigenvalues λ. This boundary case when the periodic character of the transient process ends, i.e. when $\xi \geq 1$, is determined by $R \geq 2\sqrt{I/C}$.

Finally, as presentation in the form of transfer functions has both clarity and conviction, the model given by Equations (2.2-3.21) and (2.2-3.22) should be translated into the complex region using the already mentioned transformation $\mathbf{G}(s) = \mathbf{C}\,(s\mathbf{I} - \mathbf{A})^{-1}\,\mathbf{B} + \mathbf{D}$.

Matrices $\mathbf{C}$, $\mathbf{B}$ and $\mathbf{D}$ have already been determined and only the inverse matrix of matrix $|(s\mathbf{I} - \mathbf{A})|$ is unknown.

$$(s\mathbf{I} - \mathbf{A})^{-1} = \frac{adj(s\mathbf{I}-\mathbf{A})}{det(s\mathbf{I}-\mathbf{A})} = \frac{1}{det(s\mathbf{I}-\mathbf{A})} \begin{vmatrix} s+\dfrac{R}{I} & \dfrac{1}{C} \\[2mm] -\dfrac{1}{I} & s \end{vmatrix} , \qquad (2.2\text{-}3.28)$$

$$det(s\mathbf{I}-\mathbf{A}) = s^2 + \frac{R}{I}s + \frac{1}{IC} . \qquad (2.2\text{-}3.29)$$

The final expression, after the necessary multiplications demanded by the upper transformation for obtaining $\mathbf{G}(s)$, is as follows

$$\begin{bmatrix} \Delta P_2(s) \\[3mm] \Delta M_1(s) \end{bmatrix} = \begin{bmatrix} \dfrac{1}{ICs^2+RCs+1} & \dfrac{-(Is+R)}{ICs^2+RCs+1} \\[3mm] \dfrac{Cs}{ICs^2+RCs+1} & \dfrac{1}{ICs^2+RCs+1} \end{bmatrix} \begin{bmatrix} \Delta P_1(s) \\[3mm] \Delta M_2(s) \end{bmatrix} \qquad (2.2\text{-}3.30)$$

$$\mathbf{Y}(s) = \mathbf{G}(s)\,\mathbf{U}(s)$$

This is the most general form of matrix transfer functions $\mathbf{G}(s)$ for fluid flow dynamics. From it can easily be obtained the special cases from the preceding two sections. If we neglect fluid mass inertia ($I = 0$), the upper equation gives matrix $\mathbf{G}(s)$ from Equation (2.2-3.23), into which we must insert $R_1 = R$, $R_2 = 0$, $T = 0$ to make it valid for case a) from Example 1, Section 2.2-2, on gas and steam flow. If, however, we neglect the possibility of fluid mass storage ($C = 0$) in the last equation, we get

$$\Delta P_2 = \Delta P_1 - (Is + R)\Delta M_2 , \qquad (2.2\text{-}3.31)$$

$$\Delta M_1 = \Delta M_2 = \Delta M \quad . \tag{2.2-3.32}$$

$$\Delta M(s) = \frac{\frac{1}{R}}{Ts+1} \left(\Delta P_1(s) - \Delta P_2(s) \right) \quad . \tag{2.2-3.33}$$

$$T = \frac{1}{R}$$

The last equation is completely the same as the Laplace transformation of Equation (2.2-3.20) from the section on liquid flow, where the possibility of mass storage was also neglected because we worked with ρ = const., i.e. C = 0. In the upper expression, unlike in (2.2-3.20), changes in the number of revolutions of pump Δn and in the cross-sectional area of the control valve ΔA_v are missing. These are variables that were neglected at the beginning of the present section.

The denominator of all the transfer functions $G_{i,j}$ (i = 1,2, j = 1,2) in Equation (2.2-3.30) is the same and equal to the characteristic equation of differential Equations (2.2-3.5) and (2.2-3.6), and also to Equation (2.2-3.23) for determining the eigenvalues of the system matrix $\mathbf{A}$. The well-known fact that the poles of transfer functions represent the eigenvalues of the system matrix $\mathbf{A}$ is confirmed.

Example 2 Calculation of dynamic coefficients. R, I, C, T, ω, Z

In the case of airflow through a pipe, using the data in Example 2, Section 2.2-2, determine the dynamic coefficients characterizing that process.

In Example 2 from the preceding section, gas or steam flows through a pipe of length L = 100 m and cross-sectional area A = 0.0314 m^2 . We have already determined capacitance and resistance

$$C = 3.7474 \cdot 10^{-5} \text{ kgm}^2/\text{N}, \quad R = 5533.78 \text{ (N/m}^2\text{)/(kg/s)} \quad .$$

Inertance I is not difficult to obtain from

$$I = L/A = 3184.7 \text{ m}^{-1} \quad .$$

Natural frequency is

$$\omega_n = (IC)^{-0.5} = 2.9 \text{ s}^{-1} .$$

The damping coefficient is

$$\xi = RC \ \omega_n/2 = 0.301$$

The eigenvalues of system matrix **A** and also the poles of the transfer functions are

$$\lambda_{1,2} = -0.8719 \pm 2.425 \text{ j} .$$

It is not difficult to see to which physical process the natural frequency ω_n is related, and what its real meaning is. We must first remember that the whole analysis was carried out for the case of isothermal state changes in which the velocity of sound (and that is the speed with which pressure and mass flow disturbances propagate through the pipe) is determined from the (generally known) expression

$$c_s = \sqrt{R\vartheta} = \sqrt{287 \cdot 292} = 289.5 \text{ m/s}$$

In practice fast disturbances in fluid state propagate in adiabatic conditions. This, however, does not change anything in the present analysis. As we said earlier, we then work with polytropic capacitance C_n = C/n instead of with isothermal capacitance C. In the adiabatic case for n = $\varkappa$ and for the velocity of sound we have $c_s = \sqrt{\varkappa R\vartheta}$.

The time necessary for the disturbance, propagating with the velocity of sound c_s to pass through the whole length of the pipe L, is

$$T_L = \frac{L}{c_s} = 0.345 \text{ s} . \qquad\qquad (2.2\text{-}3.34)$$

The natural frequency oscillation is equal to the inverse value of time T_L

$$\omega_n = \frac{1}{T_L} = 2.9 \text{ s}^{-1}$$

That the upper expression satisfies can also be seen from the following derivation

$$\omega_n = \frac{1}{\sqrt{IC}} = \frac{1}{\sqrt{\dfrac{L}{A}\dfrac{V}{R\vartheta}}} = \frac{\sqrt{R\vartheta}}{L} = \frac{c_s}{L} = \frac{1}{T_L} \quad . \qquad (2.2\text{-}3.35)$$

In engineering literature we very often encounter, besides the coefficients we have already introduced and used and the just-mentioned time for disturbance transfer along the pipe T_L, so-called **impedance** defined as follows for fluid transport

$$Z = \frac{c_s}{A} = \sqrt{\frac{I}{C}} \quad . \qquad (2.2\text{-}3.36)$$

It is not difficult to show that

$$I = Z\,T_L \quad . \qquad (2.2\text{-}3.37)$$

$$C = \frac{T_L}{Z} \quad . \qquad (2.2\text{-}3.38)$$

$$IC = T_L^2 \quad . \qquad (2.2\text{-}3.39)$$

Finally, to enable comparison with the case of spatially distributed processes for the conditions belonging to Figure 2.2-3.4, we give the matrix transfer functions

$$\begin{bmatrix} \Delta P_1(s) \\ \Delta M_2(s) \end{bmatrix} = \begin{bmatrix} \dfrac{1}{ICs^2 + RCs + 1} & \dfrac{Is + R}{ICs^2 + RCs + 1} \\ \dfrac{-Cs}{ICs^2 + RCs + 1} & \dfrac{1}{ICs^2 + RCs + 1} \end{bmatrix} \begin{bmatrix} \Delta P_2(s) \\ \Delta M_1(s) \end{bmatrix} \qquad (2.2\text{-}3.40)$$

2.3 HEAT PROCESSES

The field covered by this term is very wide and the variety of devices, objects and systems in which processes of heat production, transfer and accumulation occur is so great that we must limit ourselves to the basic processes we meet in technical practice. Deriving mathematical models and analyzing the dynamics of such basic processes is a step in the direction of comprehensive insight into much more complex phenomena. Like on the preceding pages, here too only processes with lumped parameters will be studied, but we will also try to show that the models obtained are in fact only rougher (which does not in any way mean bad, but very often of great use) approximations of cases in which heat properties change throughout a space volume.

Heat exchangers are one of the most frequently encountered parts of plants, and much of this chapter will treat the dynamics of heat exchange. Heat exchangers are usually devices with metal walls whose thickness is kept as small as possible, only as much as is necessary to give strength. Since metals are very conductive, in our derivations we will often assume that the thermal conductivity coefficient λ is infinite. The result of this assumption is that the temperature of the wall is the same across its whole thickness ($\vartheta_w(z)$ = const.).

The derivations for heat transfer through a wall will in most cases be valid both for plane and for curved walls of pipe exchangers. In them the wall thickness is so small compared with the pipe radius that neglecting curvature effects will not introduce essential errors into the models obtained. An exception are high-pressure, thick-walled pipes for which the above assumption does not hold.

As an example of a markedly nonlinear process we will study radiation heat transfer, and after linearization of the initial model the linear and the nonlinear time responses will be compared.

Finally, at the end of this section direct heat exchangers will be modeled. These are usually tanks (well or not-so-well insulated, and with thicker or thinner walls, depending on the pressure under which the process develops) in which several mass flows are mixed and the mixture heated by heating elements (electric, steam and so on). In the process of modeling we will begin with the most elementary example with many approximations, and by gradually discarding those initial assumptions it will be shown how the model of dynamic temperature change grows in complexity.

Example 1 Heat transfer through a wall

Figure 2.3-1 shows a process of heat transfer through a wall between fluids 1 and 2.

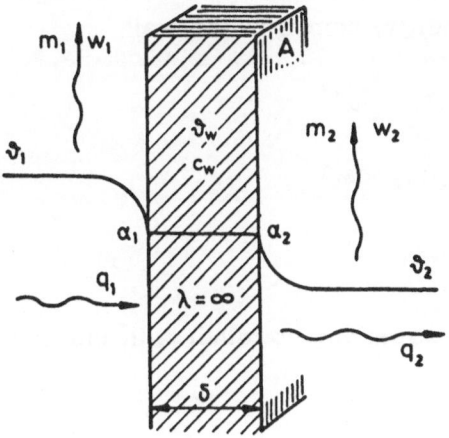

Fig. 2.3-1 Heat transfer through a wall

a) Derive a model describing the dynamics of the wall temperature depending on changes in temperatures ϑ_1 and ϑ_2. The basic assumptions are:

- the thermal conductivity coefficient through the wall $\lambda = \infty$, which means that ϑ_w is constant across the whole width of the wall,

- heat transfer on both sides is convective,

- the coefficients of convective heat transfer α_1 and α_2 are not functions of temperature,

- mass fluid flow rates m_1 and m_2 are constant (therefore, α_1 and α_2 are also constant).

b) Determine the transfer functions relating heat flow transferred onto fluid 2 with temperature.

a) The equation for the conservation of energy for a wall of surface-area A is

$$e_1 - e_2 = \frac{dE}{dt} \quad . \tag{2.3-1}$$

e_1 and e_2 are heat (energy) flow rates brought to the wall and led from it, and E heat (energy) stored in the wall.

$$e_1 = Aq_1 = A\alpha_1(\vartheta_1 - \vartheta_w) \quad , \tag{2.3-2}$$

$$e_2 = Aq_2 = A\alpha_2(\vartheta_w - \vartheta_2) \quad , \tag{2.3-3}$$

$$E = Mu = A\delta\rho c_w\vartheta_w \quad . \tag{2.3-4}$$

The last three equations, together with the first, give

$$\frac{\rho\delta c_w}{\alpha_1+\alpha_2} \frac{d\vartheta_w}{dt} + \vartheta_w = \frac{\alpha_1}{\alpha_1+\alpha_2} \vartheta_1 + \frac{\alpha_2}{\alpha_1+\alpha_2} \vartheta_2 \quad . \tag{2.3-5}$$

We have obtained a linear ODE so there is no need for linearization and introducing the symbols Δ. Equation (2.3-5), as long as the initial assumptions are true, is valid both for real temperature changes and also for changes in the deviation of those temperatures from the initial steady state.

As in the preceding cases, here we can also show that the constant beside ϑ_w has the dimension of time and, therefore, represents the time constant T of heat transfer. Its physical meaning becomes clear from the following formulas, in which the same definitions for capacitance C and resistance R are used as in earlier discussion.

Capacitance C equals the ratio of change in stored heat (energy) E to change in heat effort - temperature ϑ_w, so (2.3-4) gives

$$C = \frac{dE}{d\vartheta_w} = A\delta\rho c = Mc \tag{2.3-6}$$

Resistance R equals the ratio of change in heat effort (temperature) to change in heat flows e_1 and e_2. As resistance to heat transfer exists both on the side of fluid 1 and on the side of fluid 2, the total resistance will depend on both the individual resistances R_1 and R_2. Equations (2.3-2) and (2.3-3) give

$$R_1 = \frac{d\vartheta_w}{de_1} = -\frac{1}{A\alpha_1} \quad , \qquad\qquad (2.3\text{-}7a)$$

$$R_2 = \frac{d\vartheta_w}{de_2} = \frac{1}{A\alpha_2} \quad . \qquad\qquad (2.3\text{-}7b)$$

The fact that R_1 is negative represents the physical fact that when temperature ϑ_w increases, the heat flow rate e_1 on the wall decreases. Now we can show the meaning of constant T.

$$T = \frac{\delta\rho C}{\alpha_1+\alpha_2} = \frac{A\delta\rho C}{A\alpha_1+A\alpha_2} = \frac{C}{\frac{1}{|R_1|}+\frac{1}{R_2}} = \frac{C}{\frac{1}{R}} = RC \quad . \qquad (2.3\text{-}8)$$

Equation (2.3-5) can conveniently be shown in block diagram form and this structure of blocks is present in all first-order (proportional) systems. It is useful to compare this presentation with Figure 2.1-2.7.

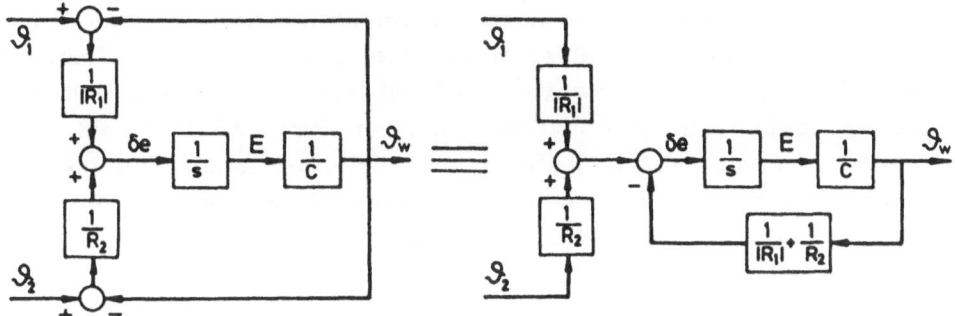

Fig. 2.3-2 Block diagram of temperature change dynamics
in pipe wall ϑ_w

b) The transfer functions relating heat flow rate e_2 to fluid 2 depending on temperature change are obtained after carrying out a Laplace transformation of Equations (2.3-3) and (2.3-5). If the expression for $\theta(s)$ from the transformed Equation (2.3-5) is inserted into the transformed Equation (2.3-3), after rearrangement we get

$$Q_2(s) = \frac{K}{T_n s+1} \Theta_1(s) - \frac{K(T_b s+1)}{T_n s+1} \Theta_2(s) \quad . \qquad (2.3\text{-}9)$$

$Q_2(s)$ is a symbol for the Laplace transformation of heat flow rate $e_2(t)$.

In Section 3.2, when the heat transfer process is observed as a process with distributed parameters, for the case of a finite λ, Equation (3.2-62) will be an approximated transfer function identical to upper Equation (2.3-9). Here we must indicate that $e_2(t)$ will have a derivative dependence on temperature change ϑ_2 of the heated fluid. A more detailed presentation of those transfer processes is given in Section 3.2.

Constants K, T_n and T_b are given by the expressions

$$K = \frac{\alpha_1 \alpha_2}{\alpha_1 + \alpha_2} A \quad , \qquad\qquad (2.3\text{-}10)$$

$$T_n = \frac{\delta \rho c}{\alpha_1 + \alpha_2} \quad , \qquad\qquad (2.3\text{-}11)$$

$$T_b = \frac{\delta \rho c}{\alpha_1} \quad . \qquad\qquad (2.3\text{-}12)$$

From Equation (2.3-8) for T we can see that this constant, besides depending on other variables, also depends on heat transfer coefficients α_1 and α_2. The determination of these coefficients is the most critical part of the calculation since the values of the other constants in (2.3-8) are not difficult to obtain. Here, however, we will not enter the problem of calculating coefficients of convective heat transfer. Whole books have been written on this subject, thermodynamic atlases, tables and the like, so for these needs we refer the reader to specialized literature. But it is, nevertheless, necessary to study some typical technical cases and see the range within which time constant T changes for such processes.

Let 80 °C temperature water flows through an uninsulated hot-water pipe whose walls are $\delta = 5$ mm thick. If the heat transfer coefficient to the surrounding air is $\alpha_1 = 10$ W/m^2 ., and to the internal surface of the pipe $\alpha_2 = 230$ W/m^2 . for $\rho = 7850$ kg/m^3 , c = 0.50 kJ/kg., the time constant T is

$$T = \frac{\delta \rho c}{\alpha_1 + \alpha_2} = \frac{7850 \cdot 0.005 \cdot 500}{230 + 10} = 81.77 \text{ s} \quad .$$

The superheater pipe of an electric plant steam generator of thickness $\delta = 0.005$, c = 500, and with coefficients $\alpha_1 = 60$, $\alpha_2 = 2400$, has the time constant

$$T = \frac{7850 \cdot 0.005 \cdot 500}{60 + 2400} = 7.98 \text{ s} \quad .$$

These two simple numerical examples clearly show that T will decrease with an increase in the heat transfer coefficients, i.e. with a decrease of the resistances to heat transfer. It must also be observed that in the case of great. differences in coefficients α, the time constant is chiefly determined by the one that is bigger. If we neglect α_1 in the second case, T becomes T = 8.177. which is only 2.5% more that the original time constant.

Up to now, on the basis of the first assumption that λ is infinite, we completely neglected the resistance to heat conduction through the wall, and T was determined only by resistance to heat convection. If boundary **conditions** in the form of forced **heat flow** are given on both boundaries of the wall, and if thermal conductivity coefficient λ is finite, in practice we use a transfer function showing the dependence of heat flow rate $e_2(t)$ on heat flow rate $e_1(t)$ on the "warmer" side of the wall

$$\frac{Q_2(s)}{Q_1(s)} = \frac{1}{T_w s + 1} \quad , \tag{2.3-13}$$

$$T_w = \frac{\rho c \delta^2}{2\lambda} \quad . \tag{2.3-14}$$

This transfer function and the time constant of the pipe T_w will be derived in Section 3.2. In the dynamic sense, the wall behaves like a first-order (proportional) system, and for a superheater steam generator pipe with λ = 46.5 W/mK

$$T_w = \frac{\rho c \delta^2}{2\lambda} = 1.055 \text{ s} \quad .$$

Finally, in the wealth of possible different cases one of the more common phenomena in pipe heat exchangers is that the fluid mass flow rates m_1 and m_2 are not constant, but can, in unsteady operation, substantially change in magnitude. In such cases the fourth assumption is no longer fulfilled and we cannot work with constant values for convective heat transfer coefficients α_1 and α_2, since they depend to a great extent on the rate of fluid flow. In practice we will, thus, very often use the following expression to determine α for single-phase fluids flowing through pipes

$$\lambda = K \, m^n \quad . \tag{2.3-15}$$

$$n = 0.8 \ldots \text{liquids}$$
$$n = 0.6 \ldots \text{gases}$$

If the fourth assumption is not fulfilled, Equation (2.3-5) is no longer linear, and if we insert into it the expressions (2.3-15) instead of α, after linearization we get the following mathematical formulation for the dynamics of wall temperature change for changes in the four possible disturbance variables

$$\frac{\delta \rho c}{\alpha_1 + \alpha_2} \frac{d\Delta\vartheta_w}{dt} + \Delta\vartheta_w = \frac{\bar{\alpha}_1}{\alpha_1 + \alpha_2} \Delta\vartheta_1 + \frac{\bar{\alpha}_2}{\alpha_1 + \alpha_2} \Delta\vartheta_2 + \frac{\bar{\vartheta}_1 - \bar{\vartheta}_w}{\alpha_1 + \alpha_2} C_1 \Delta m_1 + \frac{\bar{\vartheta}_2 - \bar{\vartheta}_w}{\alpha_1 + \alpha_2} C_2 \Delta m_2$$

$$\tag{2.3-16}$$

$$C_i = K_i n_i \bar{m}_i^{\,n_i - 1} \quad , \quad i = 1, \, 2 \quad . \tag{2.3-17}$$

If we cancel the fourth assumption, the expression for T does not change, but becomes dependent on the initial steady state and two new disturbance variables appear - mass flow rates m_1 and m_2.

================

Many more different cases could be analyzed, which will not be done here because we consider that this would not bring any essentially new discoveries. If the reader wants to gain deeper insight into the models shown here, he should turn his attention to an analysis of the results in Section 3.2. There the different phenomena are examined in more detail because of different boundary conditions on the surfaces of the walls. In this section, to continue our analysis of processes of heat transfer, we will show the typical nonlinear case of heat radiation and compare the responses of the original nonlinear model with the responses of the linear model.

================

Example 2 Radiation heat transfer

The two nearby parallel walls in Figure 2.3-3 exchange heat by radiation after, at t = 0, the temperature of wall 1 undergoes step increase from 0 °C to 85 °C. At t = 0 the temperature of wall 2 is 0 °C. The mass of the wall 2 is 0.1 kg and its specific heat c = 1000 J/kg ,. The coefficient of emission are equal and are ε_1 = ε_2 = 0.065. Both the walls are insulated towards their surroundings and only exchange heat with each other. A_1 = A_2 = 1 m^2.

a) Determine the time constant showing the speed of wall temperature increase.

b) Compare the real nonlinear response with the response of the linear model for cases of step increase in wall 1 temperature by: 8.5 °C and 85 °C.

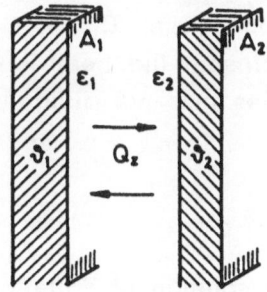

Fig. 2.3-3 Radiation heat exchange between parallel walls

a) The conservation of energy equation for wall 2 is

$$Q_z = \frac{dE_2}{dt} \quad . \tag{2.3-18}$$

$$Q_z = C_{12} \left(\vartheta_1{}^4 - \vartheta_2{}^4 \right) \frac{1}{100^4} \quad . \tag{2.3-19}$$

$$E_2 = M_2 c_2 \vartheta_2 \quad . \tag{2.3-20}$$

The three upper equations give the model demanded

$$M_2 c_2 \frac{d\vartheta_2}{dt} + \frac{C_{12}}{100^4} \vartheta_2{}^4 = \frac{C_{12}}{100^4} \vartheta_1{}^4 \tag{2.3-21}$$

The result is an ODE of pronounced nonlinearity describing the dynamics of wall 2 temperature change ϑ_2. Linearization of the upper equation gives

$$\underbrace{\frac{M_2 c_2}{4 C_{12} \dfrac{\bar{\vartheta}_2^{\,3}}{100^4}}}_{T} \frac{d\Delta\vartheta_2}{dt} + \Delta\vartheta_2 = \frac{\bar{\vartheta}_1^{\,3}}{\bar{\vartheta}_2^{\,3}} \Delta\vartheta_1 \qquad\qquad (2.3\text{-}22)$$

Time constant T, if both the numerator and the denominator are expanded by ϑ_2, has a clear physical meaning

$$T = \frac{1}{4} T_0 , \quad T_0 = \frac{M_2 c_2 \bar{\vartheta}_2}{C_{12} \dfrac{\bar{\vartheta}_2^{\,4}}{100^4}} = \frac{\text{accumulated heat}}{\text{heat flow rate}} \qquad\qquad (2.3\text{-}23)$$

If wall 1 is a black body, i.e. $C_{12} = C_2$, the expression in the denominator really does represent the heat flow rate between surfaces 1 and 2. For the given variables it is not difficult to determine that C_{12} will be $C_{12} = 0.1904$, and

$$T = \frac{0.1 \cdot 1000 \cdot 273}{4 \cdot 0.1904 \cdot 2.73^4} = 645 \text{ s} \quad .$$

b) The analytic nonlinear solution of Equation (2.3-21) has the following explicit form for time t

$$t = \frac{M_2 c_2 10^8}{C_{12} \cdot 2\vartheta_1^{\,3}} \left(\text{arctg} \frac{\vartheta_2}{\vartheta_1} - \text{arctg} \frac{\bar{\vartheta}_2}{\vartheta_1} + \frac{1}{2} \ln \frac{\dfrac{\vartheta_1 + \vartheta_2}{\vartheta_1 - \vartheta_2}}{\dfrac{\vartheta_1 + \bar{\vartheta}_2}{\vartheta_1 - \bar{\vartheta}_2}} \right) . \qquad (2.3\text{-}24)$$

The linear solution is in the already known form for a first-order system

$$\Delta\vartheta_2 = \Delta\vartheta_1 (1 - e^{\frac{1}{645} t}) \quad . \qquad\qquad (2.3\text{-}25)$$

If we use Equations (2.3-24) and (2.3-25) to calculate the responses for increases $\Delta\vartheta = 8.5 \text{ °C}$ and 85 °C , and show them graphically, we get the diagram on Figure 2.3-4.

From the following representation, which is a graphical solution of Equations (2.3-24) and (2.3-25) for the given disturbances, we can see that the difference between the nonlinear and the linear responses increases with an increase of the disturbance value.

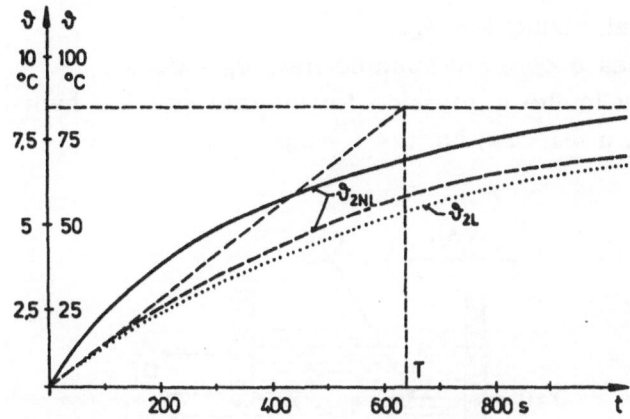

Fig. 2.3-4 Response of the nonlinear and the linear model
in radiation heat exchange

—— NL $\Delta\vartheta_1$ = 85 °C   ----- NL $\Delta\vartheta_1$ = 8.5 °C
······· L   ········ L

The ordinate has values for both the disturbances and as the scale is the same (only 10 times larger) one curve is sufficient for both cases of linear model response. For the nonlinear model, however, we get two different curves and, as we expected, the difference from the linear response is greater for the greater deviation $\Delta\vartheta_1$ = 85 °C from the steady state.

Example 3 Direct heat exchanger

In direct exchangers, heat is transferred by the direct mixing of several mass flows of different temperature. In the tank itself additional heating also takes place by electric, steam or other heaters. Figure 2.3-5 shows one such exchanger, and the sign for a mixer symbolized the complete stirring of the liquid in the tank and a homogeneous heat field, i.e. it is considered that the temperature is the same in the whole volume of the exchanger and that liquid of that temperature flows out of the tank. Derive a model of dynamic temperature change at the tank outlet with the following assumptions:

- inlet and outlet mass flow rates are constant, i.e. $m_i = m_o = m$,

- the tank is insulated,
- there is ideal mixing $\vartheta = \vartheta_0$,
- the liquid has a constant specific heat c_p = const.,
- the term P v in the expression for internal specific heat is neglected for the liquid, $u = i - Pv$, i.e. $u = i = c_p \vartheta$.

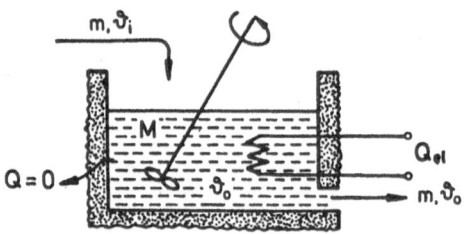

Fig. 2.3-5 Insulated direct heat exchanger

The equation for the conservation of heat has the usual form

$$e_i + Q_{el} - e_o = \frac{dE}{dt} \quad . \tag{2.3-26}$$

$$e_i = mc_p \, \vartheta_i \quad . \tag{2.3-27}$$

$$e_o = mc_p \, \vartheta_o \quad . \tag{2.3-28}$$

$$E = Mc_p \vartheta_o \quad . \tag{2.3-29}$$

When we arrange the last expression, we get

$$T\frac{d\vartheta_o}{dt} + \vartheta_o = \vartheta_i + \frac{1}{mc_p} \, Q_{el} \quad . \tag{2.3-30}$$

Time constant T = M/m, and if both the denominator and the numerator are expanded, we get

$$T = \frac{Mc_p\vartheta_o}{mc_p\vartheta_o} = \frac{\text{stored (accumulated) heat}}{\text{heat flow rate}} \quad . \tag{2.3-31}$$

The upper assumptions give a standard linear ODE for a first-order (proportional) processes. If m = const. is not satisfied, the equation is no longer linear because the products $m\vartheta_o c_p$ and $m\vartheta_i c_p$ are no longer linear expressions.

To enable comparison with the approximated transfer functions in Section 3.2, Equation (2.3-30) will be transformed into

$$\Theta_o(s) = \frac{1}{\frac{M}{m}s+1}\Theta_i(s) + \frac{\frac{1}{mc_p}}{\frac{M}{m}s+1}Q_{el}(s) \quad . \tag{2.3-32}$$

The following lines will show how "insignificant" changes in assumptions change essentially the mathematical model, thus showing all the importance of a careful choice of simplifications when the model is being formulated.

Derive a dynamic model using all the initial assumptions except that the second one is changed and two new ones are added
  - the tank is not insulated,
  - the accumulation of the tank wall is neglected (a thin-walled tank),
  - the coefficient of thermal conductivity through the wall is infinite.

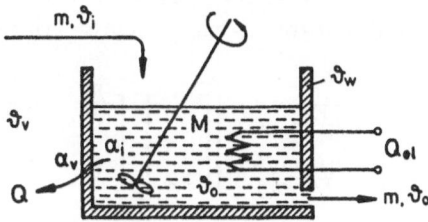

Fig. 2.3-6 Uninsulated direct heat exchanger

The equation for the conservation of heat has the following form

$$mc_p\vartheta_i + Q_{el} - \alpha_v A_v(\vartheta_o - \vartheta_v) - mc_p\vartheta_o = Mc_p\frac{d\vartheta_o}{dt} \quad . \tag{2.3-33}$$

After arrangement we get

$$T\frac{d\vartheta_o}{dt} + \vartheta_o = \frac{mc_p}{mc_p+\alpha_v A_v}\vartheta_i + \frac{1}{mc_p+\alpha_v A_v}Q_{el} + \frac{\alpha_v A_v}{mc_p+\alpha_v A_v}\vartheta_v \quad , \tag{2.3-34}$$

$$T = \frac{Mc_p}{mc_p+\alpha_v A_v} \quad . \tag{2.3-35}$$

Again this is the dynamics of the usual first-order (proportional) system, except that the constant T has changed.

If we leave out the assumption about the negligible accumulation of the exchanger wall (and keep all the other assumptions), we must formulate **two balance equations** since we now have two possible heat tanks - the liquid in the tank and the wall of the heat exchanger itself:

- liquid

$$mc_p\vartheta_i + Q_{el} - \alpha_i A_i(\vartheta_0 - \vartheta_w) - mc_p\vartheta_0 = Mc_p\frac{d\vartheta_0}{dt} \qquad (2.3-36)$$

- wall

$$\alpha_i A_i(\vartheta_0 - \vartheta_w) - \alpha_v A_v(\vartheta_w - \vartheta_v) = M_w c_w \frac{d\vartheta_w}{dt} \qquad (2.3-37)$$

As long as the convective heat transfer coefficients α_v and α_i do not depend on temperature, the upper system of equations is linear and can be written directly in the form of matrix state space equations

$$\begin{bmatrix} \vartheta_0 \\ \vartheta_w \end{bmatrix}' = \begin{bmatrix} -\dfrac{mc_p + \alpha_i A_i}{Mc_p} & \dfrac{\alpha_i A_i}{Mc_p} \\ \dfrac{\alpha_i A_i}{M_w c_w} & -\dfrac{\alpha_i A_i + \alpha_v A_v}{M_w c_w} \end{bmatrix} \begin{bmatrix} \vartheta_0 \\ \vartheta_w \end{bmatrix} + \begin{bmatrix} \dfrac{m}{M} & \dfrac{1}{Mc_p} & 0 \\ 0 & 0 & \dfrac{\alpha_v A_v}{M_w c_w} \end{bmatrix} \begin{bmatrix} \vartheta_i \\ Q_{el} \\ \vartheta_v \end{bmatrix}$$

$$(2.3-38)$$

To conclude our analysis of the dynamics of direct heat exchangers, we will now show a model which assumes that liquid stirring is very intense. This insures a **large heat transfer coefficient** α_i on the interior surface of the wall. $\vartheta_w = \vartheta_0$, and with $\lambda = \infty$ we assume that there is only one equation for the conservation of heat

$$mc_p\vartheta_i + Q_{el} - mc_p\vartheta_0 = (Mc_p + M_w c_w)\frac{d\vartheta_0}{dt} \qquad . \qquad (2.3-39)$$

After the Laplace transformation we get the following form

$$\Theta_0(s) = \underbrace{\frac{1}{\dfrac{Mc_p + M_w c_w}{mc_p}s + 1}}_{G(s)} \left(\Theta_i(s) + \frac{1}{mc_p}Q_{el}(s)\right) \qquad . \qquad (2.3-40)$$

It is useful to observe that the transfer function obtained G(s) is the same as the approximated transfer function given in Equation (3.2-27) in the section on treating the direct heat exchanger with thick walls as a system with distributed parameters.

───────────────

At the end of our analysis of the dynamics of basic heat processes we must stress that temperature changes are described as a rule by the already known equations for proportional systems. Time constant T is here also equal to the product of thermal capacitance C and resistance R, $T = CR$. From this aspect heat processes, which are in the physical sense reduced to processes of heat storage, are dynamically similar to processes of mass storage in tanks, and the main difference between these and flow processes is that heat processes have no inertia. Thus there will be no periodic, oscillatory state changes in their case.

2.4 MECHANICAL PROCESSES

The motion dynamics of the rigid body, mass point (particle) or more complex mechanisms is a classical technical science and a discipline that is studied in many institutions. Today there is a very varied selection of textbooks, handbooks, collections of examples and books from that field in the world, which show both basic and also very elaborate methods and modeling procedures. As a rule, all these books contain numerous examples (from the simplest to very complex mechanisms) and are highly specialized, so it must immediately be emphasized that this section of our book is not in any way intended as a supplement, substitute or perhaps competition to such books and collections. Its basic purpose, like that of the preceding sections, is to describe the dynamics of mechanical processes of the rigid body motion in mathematical forms usual in this book, and to thus show unity in the dynamics of these and other processes. With that in mind, the dynamic coefficients of resistance, capacitance and inertance will be redefined and their physical meaning given.

In this way we will deviate to a certain extent from the usual form of mathematical models in classical dynamics to which some readers are probably more than accustomed, but all the examples will be presented in the (here widely used) form of matrix state-space equations or, simply, by systems of first-order DE. This notation has shown itself to be of great use in designing control procedures for modern robot mechanisms and manipulators. For that purpose, namely, to synthesize the whole controlling assembly - from sensors and logical devices that are built of microprocessors or small computer units, to actuators (electrical, hydraulic or pneumatic motors) - it is essential to have a model in the very form that is used in modern control theory, and that is notation in the form of state-space equations.

Three basic procedures have become standard in obtaining models for the dynamics of the rigid body motion in mechanics:

- the use of Newton's law for the conservation of momentum

- the use of d'Alembert's principle,

- deriving Lagrange's equations of motion.

Here we must say that d'Alembert's principle is in fact only a skilful transformation of Newton's law, while Lagrange's equations of motion are based on considerations concerning the conservation of energy. It can be shown that Lagrange's equations can be derived from Newton's laws, and without carrying out a critical comparison of the methods, we will only add that in deriving a model for the dynamics of motion of more complicated systems it is advisable to do so by the simultaneous use of Lagrange's equations and of Newton's laws. Both methods must give the same result.

Before beginning the analysis of specific examples, we must point to the analogy between the two basic forms of motion in mechanics - the similarities between translation and rotation. Without any more extensive explanation, these analogies are given in the following table.

Table 2.4-1

Translation		Rotation	
z	displacement	angular displacement	φ
$w = z'$	velocity	angular velocity	$\omega = \varphi'$
$a = w' = z''$	acceleration	angular acceleration	$\varepsilon = \omega' = \varphi''$
M	mass	moment of inertia	J
F	force	torque	M_z
$I_m = Mw$	momentum	angular momentum	$L_z = J\omega$
$\dfrac{d(Mw)}{dt} = F$		$\dfrac{d(J\omega)}{dt} = M_z$	
$T = \frac{1}{2}Mw^2$	kinetic energy		$T = \frac{1}{2}J\omega^2$
$W = \int F\,dz$	work		$W = \int M_z\,d\varphi$
$P = Fw$	power		$P = M_z\omega$

In mechanical systems there is duality in the selection of the variables of effort (potential), flow and stored variables, and thus also differences in the physical meaning of dynamic coefficients. This duality results from whether we select force or velocity in translation (torque or angular velocity in rotation) for the variable of effort, and it will be shown on an example of a simple mechanical system: mass-spring-damper.

We will begin this chapter by describing the motion dynamics of a body using the simplest example of translation and the usual mathematical tools. Everything that is done in further examples for cases of translation can, without any reservations, be applied to rotation as well, using the above analogy table.

Before setting up the equations it must be pointed out that this section will in most cases analyze linear processes. This means that the variable changes will be considered small enough to make nonlinear interrelations unnecessary. It follows that there will be no linearization, which also means that there will be no deviation notation Δ beside linearized variables. In cases where we do carry out linearization, however, previously used symbols will be retained but then it must not be forgotten that the original model was in fact nonlinear.

=====

Example 1 Translation dynamics

Consider a body of mass M, under the influence of force F(t), in translational motion on a flat surface. In the general case the following forces resist motion: the force of dry (sliding) friction F_2, a force proportional to the velocity of the body (the force of viscous friction) F_3 and a force proportional to the body's displacement F_4. Derive a model that describes the dynamics of displacement z and velocity w if at t = 0 there was:

1st IC z = 0

2nd IC $\dfrac{dz}{dt}$ = w = w_0 (IC ... initial condition)

and for the following cases:

a) ideal frictionless motion,
b) motion with the presence of sliding friction,
c) motion with the presence of a resistance force proportional to the velocity w,
d) resistance force to motion proportional to body displacement z.

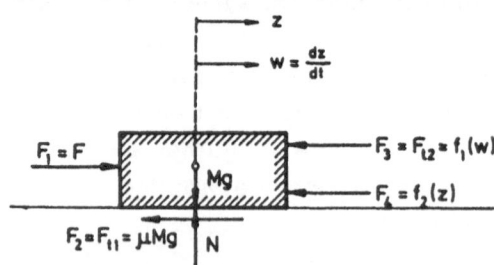

Fig. 2.4-1 A body in motion with driving and resistance forces

For each case the model will be obtained from the following basic expression describing Newton's law for the conservation of momentum

$$\sum_{i=1}^{4} F_i = \frac{d(Mw)}{dt} = M\frac{dw}{dt} \quad . \tag{2.4-1}$$

a) in the case of ideal motion without any forces, i.e. $F_i = 0$ ($i=1,4$), (2.4-1) becomes.

$$M\frac{dw}{dt} = 0 \quad . \tag{2.4-2}$$

This is an equation with a "pure" integral term for velocity w change, which with the given initial conditions becomes

$$w = w_0 = \text{const.}$$

Since the generally-known relation between velocity w and displacement z is (also an integral member without time lag)

$$\frac{dz}{dt} = w \quad . \tag{2.4-3}$$

it follows that z will continuously increase to infinity, i.e.

$$z = w_0 t \quad .$$

b) There is a resistance force of dry (sliding, Coulomb) friction, so that (2.4-1) becomes

$$M\frac{dw}{dt} = -\mu Mg \quad . \tag{2.4-4}$$

A system dynamically equivalent to this example of translation is given by Equation (2.2-1.2) for the case when the input pump ceases operation, i.e. for $m = 0$. Like liquid level H in that example, here velocity linearly decreases to zero. It can easily be shown that the following expression is the solution of (2.4-4) for the given IC

$$w = \frac{dz}{dt} = w_0 - \mu gt \quad . \tag{2.4-5}$$

Equation (2.4-5) gives the analytical expression for displacement z

$$z = w_0 t - \frac{\mu g}{2} t^2 \quad . \tag{2.4-6}$$

This process of free translation with the presence of dry friction can also be shown in block diagram form.

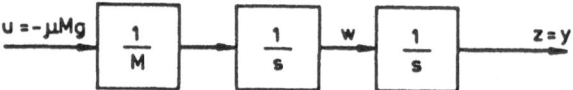

Fig. 2.4-2 Block diagram of translation with dry friction

If we select displacement $z = x_1$ and velocity $w = x_2$ for state variables, the system of Equations (2.4-3) and (2.4-4) can easily be given in the form of matrix state-space equations

$$\begin{bmatrix} z \\ w \end{bmatrix}' = \begin{bmatrix} 0 & 1 \\ 0 & 0 \end{bmatrix} \begin{bmatrix} z \\ w \end{bmatrix} + \begin{bmatrix} 0 \\ \frac{1}{M} \end{bmatrix} \begin{bmatrix} -\mu Mg \end{bmatrix} \quad . \tag{2.4-7}$$

X' = A X + B U

The eigenvalues of matrix **A** are equal to zero, $\lambda_1 = \lambda_2 = 0$, indicating the presence of the two integral terms without a time lag shown on Figure 2.4-2.

c) Driving force F and a force of resistance proportional to velocity act on the body

$$F_3 = - Dw \tag{2.4-8}$$

D is the coefficient of that linear dependence and is also called the coefficient of viscous friction. Now (2.4-1) becomes

$$M\frac{dw}{dt} + Dw = F \quad . \tag{2.4-9}$$

This equation is the classical and already several times obtained example of a first-order proportional system for the dynamics of a body's velocity w, in which time constant T equals

$$T = \frac{M}{D} \quad . \tag{2.4-10}$$

At initial moment t = 0 the velocity is w_0 and if the numerator and denominator are expanded by that velocity, the physical meaning of the constant T becomes clearer.

$$T = \frac{Mw_0}{Dw_0} = \frac{I_m}{F} = \frac{momentum}{force} \qquad (2.4\text{-}11)$$

T is thus the ratio of the total momentum stored to the force that acts on the body in the steady state observed. If displacement z is important, it can, as in the preceding case, be obtained by the simple integration of velocity w.

If (2.4-9) is compared with the other equations of proportional first-order systems in preceding sections, it can be seen that velocity w is the variable of effort or potential, and force F the variable of flow. The stored variable is momentum $I_m = Mw$. In this case the capacitance and resistance are

$$C = \frac{AV}{E} = \frac{Mw}{w} = M \quad , \qquad (2.4\text{-}12)$$

$$R = \frac{E}{F} = \frac{w}{F} = \frac{1}{D} \quad . \qquad (2.4\text{-}13)$$

and the time constant, as until now, is

$$T = RC = \frac{M}{D}$$

If coefficients C and R that have just been introduced are used, the following block diagram shows translational motion in the case of a resistance force proportional to velocity.

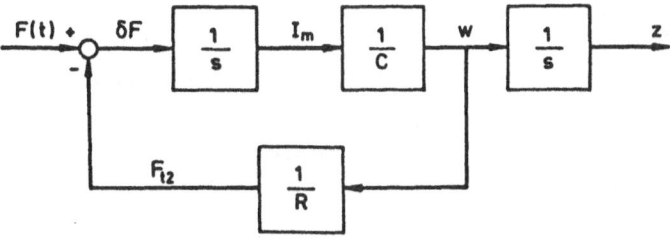

Fig. 2.4-3 Block diagram of translational motion if the resistance force is linearly dependent on velocity

The motion described by two OLDE of the first order (2.4-3) and (2.4-9) can be given as follows by state equations

$$
\begin{bmatrix} z \\ w \end{bmatrix}' = \begin{bmatrix} 0 & 1 \\ 0 & -\dfrac{D}{M} \end{bmatrix} \begin{bmatrix} z \\ w \end{bmatrix} + \begin{bmatrix} 0 \\ \dfrac{1}{M} \end{bmatrix} \begin{bmatrix} F \end{bmatrix} \qquad . \tag{2.4-14}
$$

X' = A X + B U

It is not difficult to solve the characteristic equation to obtain the eigenvalues of matrix **A**. They differ from each other and are

$$\lambda_1 = 0 \qquad ,$$

$$\lambda_2 = -\frac{D}{M} = -\frac{1}{T}$$

The first eigenvalue shows the integral character of the process of position change in time, i.e. displacement z. The second shows a stable transient process of a first-order (proportional) system describing velocity w change due to changes in force F acting on the body.

If we substitute (2.4-3) into (2.4-9), the model is transformed into the following OLDE of the second order, which is the usual mathematical presentation of the dynamics of translational motion and shows displacement z

$$
M \frac{d^2 z}{dt^2} + D \frac{dz}{dt} = F \qquad . \tag{2.4-15}
$$

This way of formulating the model with second-order derivatives of the displacement variable z is usual and widespread in classical mechanics. There is, nevertheless, some similarity in the different ways in which models are formulated by the state-space method and in classical mechanics. It lies in the fact that both approaches try to obtain a system comprising DE, only in mechanics it is usual for them to be DE of the second, and in the state-space method DE of the first order.

The method we used up to now to present the dynamics of a body in translational motion is in a way unusual in classical mechanics. Therefore, this last case will be presented in the manner usual in that discipline.

As a rule, three standard elements are used to represent mechanical structures. With their help it is possible to combine and represent dynamically even the most complex mechanisms, as has been shown, to a satisfactory degree. They are: an inertia element of mass M, a spring of constant c and a damper of coefficient of viscous friction D. (Sometimes and particularly in anglosaxon literature the spring constant is denoted with S, but in these lines we will retain the standard symbol usual in mechanics - c.) The basic properties of each of these elements are represented on the following figures and given with the expressions beside them.

Spring $F = cz = c(z_1 - z_2)$ (2.4-16)

Damper $F = D(w_1 - w_2)$ (2.4-17)
                                                  $= D(z'_1 - z'_2)$

Mass $F = Ma = M\dfrac{d^2z}{dt^2}$ (2.4-18)

All the elements shown here are also present in the case of rotation, according to the analogies given in Table 2.4-1.

Torsional spring $M_z = c_1(\varphi_1 - \varphi_2)$
                                                   (2.4-19)

Torsional damper $M_z = D_t(\omega_1 - \omega_2)$ (2.4-20)
                                              $= D_t(\varphi'_1 - \varphi'_2)$

Body with moment of inertia J $M_z = J\varepsilon = J\dfrac{d^2\varphi}{dt^2}$
                                                     (2.4-21)

(c_t .... torsional spring constant, D_t .... torsional coefficient of viscous friction).

The above elements are linear approximations of real mechanical objects and are as a rule valid for small displacements. In the case of the spring the approximation is correct if we do not cross the proportional limit. The viscous damper is a good substitute for energy loss and resistance forces only in the region of real linear dependence of resistance force to motion on velocity. Since this is not a book whose purpose is to examine the validity of introducing these linear elements into an analysis of mechanical systems, nor to study the range within which such linear approximations are satisfactory, these elements will in the further text be used in the way usual in mechanics, taking into account the fact that their application in fact already represents linearization of the task. The last case can thus be shown by the following combination of elements of inertia and damping - Fig. 2.4-4.

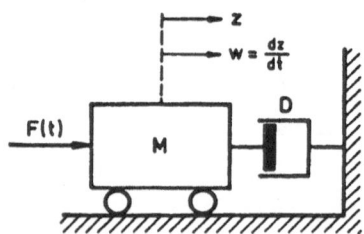

Fig. 2.4-4 Translation shown as a combination of elements
of inertia and damping

d) Driving force F and a resistance force proportional to displacement z act on the body.

$$F_4 = cz \quad .$$

This case, when resistance is proportional to displacement z, is shown by the simplest element, a spring, and the following figure represents this type of translation.

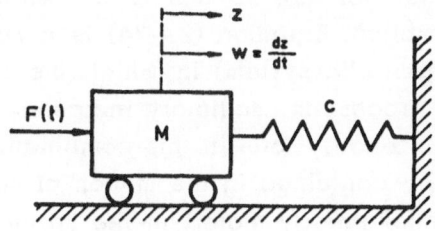

Fig. 2.4-5 Translation shown as a combination of elements
of inertia and spring

For small displacements where the principle of linearity is valid,
Equation (2.4-1) becomes

$$F_1 - F_4 = M\frac{dw}{dt} \quad . \tag{2.4-22}$$

i.e.

$$M\frac{dw}{dt} + cz = F \quad . \tag{2.4-23}$$

The model of motion now comprises Equations (2.4-3) and (2.4-23), with
whose help it is easy to show that

$$M\frac{d^2z}{dt^2} + cz = F \quad . \tag{2.4-24}$$

It is not difficult to combine these two OLDE of the first order into
matrix state-space form

$$\begin{bmatrix} z \\ w \end{bmatrix}' = \underbrace{\begin{bmatrix} 0 & 1 \\ -\frac{c}{M} & 0 \end{bmatrix}}_{\mathbf{A}} \begin{bmatrix} z \\ w \end{bmatrix} + \begin{bmatrix} 0 \\ \frac{1}{M} \end{bmatrix} \begin{bmatrix} F \end{bmatrix} \quad . \tag{2.4-25}$$

The eigenvalues of matrix **A** are now a complex conjugate pair

$$\lambda_{1,2} = \pm\sqrt{\frac{c}{M}} \, j \quad , \tag{2.4-26}$$

which shows that the transient process will be periodic. The real part equals zero (which is a direct consequence of the assumption that there are no resistance friction forces), so this is the already known case of undamped oscillatory motion. Equation (2.4-24) is a DE of a second-order (proportional) system (periodic system) in which the damping part equals zero. The fact that the process is oscillatory indicates that there must be a possibility for change in energy form in this combination. Intuitively, this is clear. The kinetic energy contained in the motion of mass M by velocity w changes into the potential energy stored in the spring by its compression and stretching, i.e. by changes in displacement variable z. However, the dynamic coefficient, inertance I, must also be found. This inertance is part of the oscillatory process because, as we already showed in Section 2.2, such processes are possible only if there is a link between elements of capacitance and inertance. The path leading to that inertance I will, nevertheless, not be intuitive since it is more suitable to use mathematical expressions and definitions from which it follows directly

$$I = \frac{E}{\dfrac{dF}{dt}} = \frac{Edt}{dF} = \frac{wdt}{dF} = \frac{dz}{dF} = \frac{1}{c} \quad . \tag{2.4-27}$$

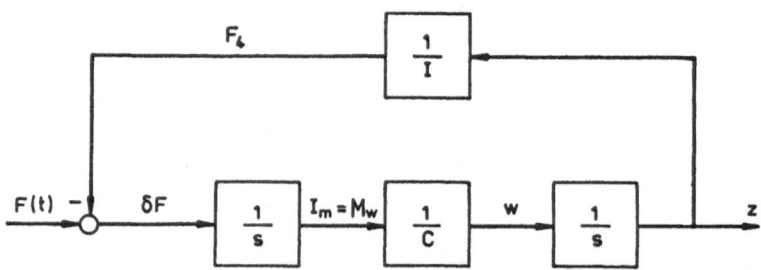

Fig. 2.4-6 Block diagram of translational motion of a body if resistance depends linearly on displacement

We have, thus, used the definition by which I equals the ratio of effort to gradient of flow, and a little calculation gave us that coefficient. The dynamic process of change in position z due to force F is given in the block diagram as shown in Figure 2.4-6. Coefficients C and I are given by (2.4-12) and (2.4-27) .

The preceding example has brought us closer to considering as one whole the dynamics of mechanical and other processes. Examples have been presented in the form of state-space equations and block diagrams, coefficients of capacity, resistance and inertia have been given, and in this way we deviated from the usual analytical tools of classical mechanics.

The following example will show the earlier mentioned duality which appears in mechanics as a result of the selection of dynamic variables and coefficients. A similar duality also exists in electrical circuits.

Example 2 Duality of dynamic variables and coefficients in mechanics

Figure 2.4-7 shows two mass-spring-damper systems. Derive DE describing their dynamics, determine their dynamic variables and dynamic coefficients, and draw their block diagrams. The dry friction resistance is neglected, i.e. $F_2 = 0$.

The systems shown in Figure 2.4-7 are similar in dynamic characteristics. The essential difference is that in Figure a) the displacements z, i.e. velocities $w = z'$ are equal, which does not hold for Figure b). On that Figure the velocities are different but the forces are equal.

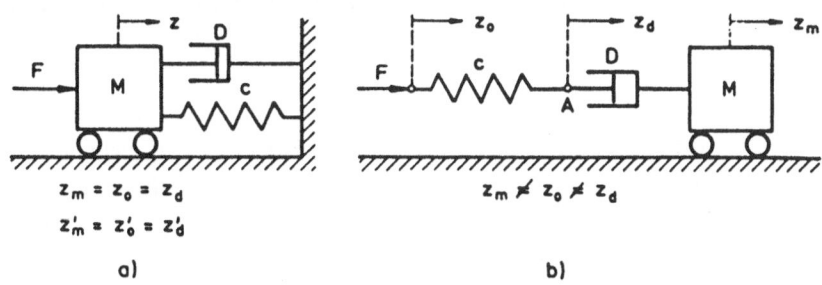

Fig. 2.4-7 Mechanical systems with:
a) equal velocities
b) equal forces

Applying Newton's law for the conservation of momentum to the system on Figure 2.4-7a (where $z_m = z_0 = z_d = z$) yields

$$\underbrace{F}_{F_1} - \underbrace{cz}_{F_4} - \underbrace{D\frac{dz}{dt}}_{F_3} = M\frac{d^2z}{dt^2} \quad , \tag{2.4-28}$$

i.e.

$$M\frac{d^2z}{dt^2} + D\frac{dz}{dt} + cz = F \quad . \tag{2.4-29}$$

In the system on Figure 2.4-7b the velocities, i.e. the displacements, are no longer equal. Here the forces are equal. If we derive d'Alembert's equations for the equilibrium of forces for point A and for mass M, we get the following two equations

$$c(z_0 - z_d) = D(z'_d - z'_m) \quad . \tag{2.4-30}$$

$$D(z'_d - z'_m) = Mz''_m \quad . \tag{2.4-31}$$

Rearrangement yields

$$\frac{M}{c}w''_m + \frac{M}{D}w'_m + w_m = w_0 \quad . \tag{2.4-32}$$

The case on figure 2.4-7a, described by (2.4-29), has already been analyzed in the preceding example and mathematically described by (2.4-24), only here the resistance force Dw, represented by a damper, is also included. The dynamic variables and coefficients defined earlier also satisfy in this case. Now they must be determined for the system on Figure 2.4-7b.

It has already been mentioned that force F is a common and equal variable for all the three elements, and here it is the variable of effort. The variable of flow, also an input variable, is now velocity w. This is the change from the preceding example, where the dynamic variables were the opposite - the variable of effort was velocity w and of flow force F. The stored variable is the variable of spring stretching and compression $\delta z = z_0 - z_d$.

The dynamic coefficients (capacitance, resistance and inertance) are obtained directly from their definitions

$$C = \frac{AV}{E} = \frac{z}{F} = \frac{1}{c} \quad , \tag{2.4-33}$$

$$R = \frac{E}{F} = \frac{F}{w} = D \quad , \tag{2.4-34}$$

$$I = \frac{E}{\frac{dF}{dt}} = \frac{F}{\frac{dw}{dt}} = \frac{F}{z''} = M \quad . \tag{2.4-35}$$

The Table 2.4-2 shows the dynamic variables and coefficients, i.e. their possible duality in mechanics (the case of rotational motion is also presented).

Table 2.4-2

	Translation		Rotation	
AV F E	$I_m = Mw$ F w	z w F	$L_z = J\omega$ M_z ω	φ ω M_z
C	M	$\frac{1}{c}$	J	$\frac{1}{c_t}$
R	$\frac{1}{D}$	D	$\frac{1}{D_t}$	D_t
I	$\frac{1}{c}$	M	$\frac{1}{c_t}$	J

Figure 2.4-8 represents a block diagram for the descriptions given by Equations (2.4-29) and (2.4-32).

The example in which force F is the flow variable is more natural and easier to understand, and this is the case for which we will derive state-space equations and analyze eigenvalues. If state variables and inputs are selected as follows

$x_1 = z$
$x_2 = w = z'$
$u = F \quad ,$
two first-order OLDE are obtained

$x_1' = z' = w = x_2$

$$x_2' = z'' = -\frac{c}{M}z - \frac{D}{M}z' + \frac{1}{M}F = -\frac{c}{M}x_1 - \frac{D}{M}x_2 + \frac{1}{M}F$$

which it is easy to set down in the form of matrix state-space equations

$$\begin{bmatrix} z \\ w \end{bmatrix}' = \begin{bmatrix} 0 & 1 \\ -\frac{c}{M} & -\frac{D}{M} \end{bmatrix} \begin{bmatrix} z \\ w \end{bmatrix} + \begin{bmatrix} 0 \\ \frac{1}{M} \end{bmatrix} \begin{bmatrix} F \end{bmatrix} \quad . \tag{2.4-36}$$

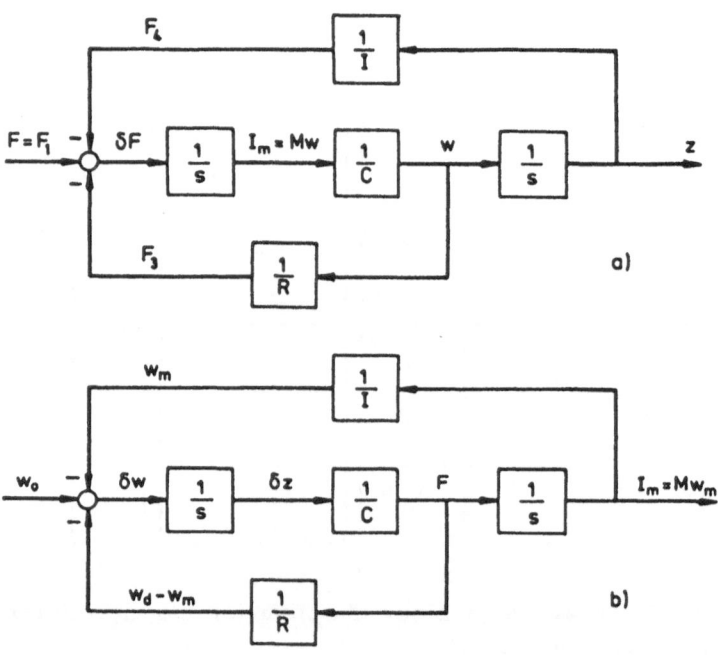

Fig. 2.4-8 Block diagrams of the mechanical system
mass-spring-damper

It is not difficult to establish that (2.4-7), (2.4-14) and (2.4-25) are only special cases of the upper equations. The eigenvalues are obtained by solving Equation $(\lambda I - A) = 0$ so that

$$\lambda_{1,2} = \frac{-D \pm \sqrt{D^2 - 4cM}}{2M} \quad . \tag{2.4-37}$$

An analysis of the last expression points to three possible sets of values for the eigenvalues λ:

a) $D^2 < 4$ cM, $\lambda_{1,2}$ is a pair of complex conjugate values with a negative real part, the response is of damped periodic character,

b) $D^2 = 4$ cM, $\lambda_1 = \lambda_2$ is a pair of equal negative real roots, the response is unperiodic,

c) $D^2 > 4$ cM, $\lambda_1 \neq \lambda_2$, the roots are real, negative and different, the response is unperiodic and strongly damped.

The following standard coefficients for periodic processes are very often used in practice.

The natural frequency of undamped (D = 0) oscillation

$$\omega_n = \sqrt{\frac{c}{M}} = \sqrt{\frac{1}{IC}} \tag{2.4-38}$$

The damping ratio ξ

$$\xi = \frac{\text{actual damping coefficient}}{\text{critical damping coefficient}} = \frac{D}{2\sqrt{cM}} \tag{2.4-39}$$

Substituting these variables into (2.4-27) for eigenvalues

$$\lambda_{1,2} = -\xi\omega_n \pm \omega_n\sqrt{\xi^2 - 1} = \sigma \pm \omega j \tag{2.4-40}$$

The analysis that was carried out for Equation (2.2-3.27) is valid here also, except that in this case undamped oscillations (D = 0), and (2.4-13) makes resistance R infinite. In the mentioned analysis we must, therefore, take the coefficient of viscous friction D instead of R in such cases.

========

Example 3 Dynamic models of different mechanical systems

Figure 2.4-9 shows three simple mechanical systems with given input and output variables. Derive a dynamic model for those systems and determine the eigenvalues of system matrix **A**. The basic assumptions for the individual systems are:

a) the cylinder rolls without sliding,

b) the driving torque is transferred, without sliding, from drum 1 to drum 2, which lifts a load by winding a stiff, unstretchable rope,

c) the mass and moment of inertia of the lever are neglected.

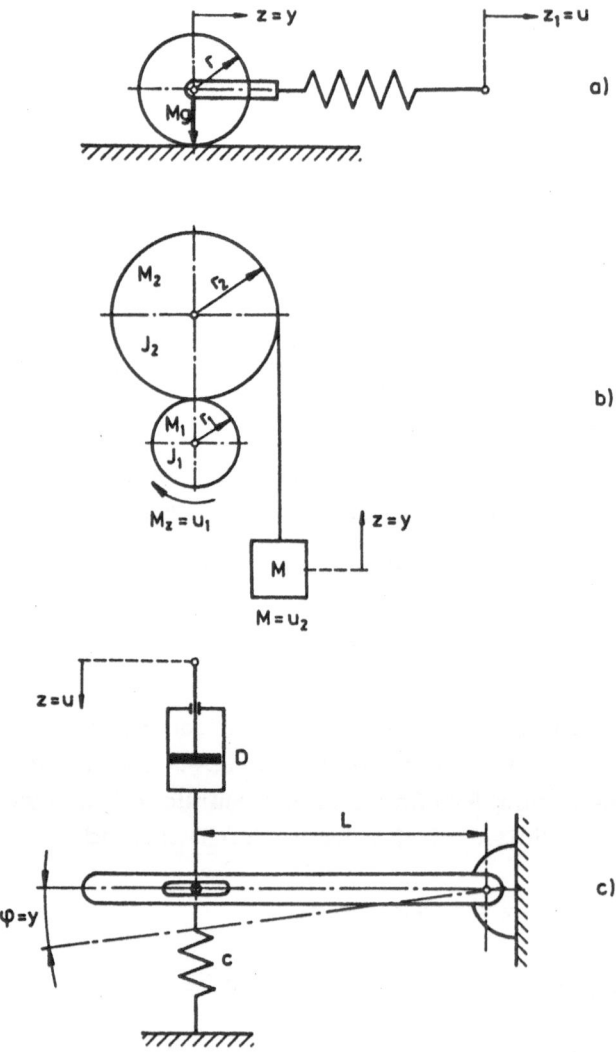

Fig. 2.4-9 Three simple mechanical systems

a) The equation describing relations between displacement z of the cylinder and displacement z_1 of the end of the spring are obtained by deriving the equations of equilibrium for horizontal forces and torques around the centre of the cylinder (d'Alembert's principle) with the help of Figure 2.4-10.

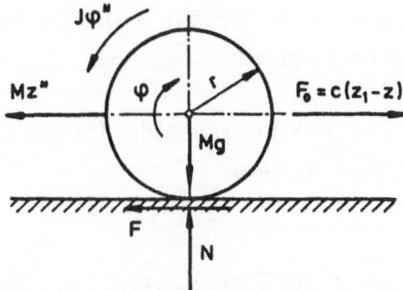

Fig. 2.4-10 The forces acting on the cylinder

The equilibrium of forces yields

$$Mz" = F_0 - F = c(z_1 - z) - F \quad . \tag{2.4-41}$$

and the equilibrium of torques around the cylinder centre yields

$$J\varphi" = Fr \quad . \tag{2.4-42}$$

Since there is no sliding $z = r\varphi$, so (2.4-42) becomes

$$Jz" = Fr^2 \quad . \tag{2.4-43}$$

For the cylinder $J = Mr^2/2$, so if F is expressed from the last equation, (2.4-41) becomes

$$z" + \frac{2}{3}\frac{c}{M}z = \frac{2}{3}\frac{c}{M}z_1 \tag{2.4-44}$$

With the state variables $x_1 = z$ and $x_2 = w$, it follows that

$$\begin{bmatrix} z \\ w \end{bmatrix}' = \begin{bmatrix} 0 & 1 \\ -\dfrac{2c}{3M} & 0 \end{bmatrix} \begin{bmatrix} z \\ w \end{bmatrix} + \begin{bmatrix} 0 \\ \dfrac{2c}{3M} \end{bmatrix} [z_1] \tag{2.4-45}$$

The eigenvalues of matrix **A** are a complex conjugate pair

$$\lambda_{1,2} = \pm \sqrt{\frac{2c}{3M}} \cdot j \tag{2.4-46}$$

which shows that this is a typical periodic undamped process of a second-order (proportional) system.

b) Figure 2.4-11 shows all the possible forces and torques on each of the parts of the load-lifting system.

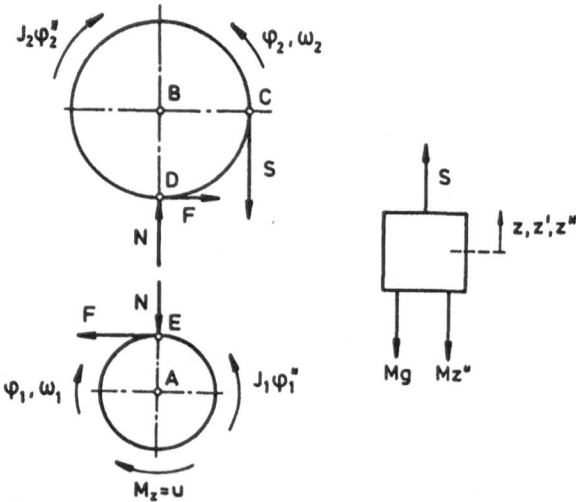

Fig. 2.4-11 Forces and torques on a load-lifting system

The equations for torque equilibrium around the centres of rotation A and B and for the equalities of the vertical forces acting on mass M are

$$J_1 \varphi_1'' = M_z - Fr_1 \ . \tag{2.4-47}$$

$$J_2 \varphi_2'' = Fr_2 - Sr_2 \ . \tag{2.4-48}$$

$$Mz'' = S - Mg \tag{2.4-49}$$

Point C of drum 2, assuming the rope is stiff and unstretchable, moves at the same velocity as the raising load $w = z'$, and since there is no sliding, the velocities of points E and D will be equal. The following relations are thus fulfilled

$$I_2\omega_2 = z' \qquad\qquad I_2\omega_2' = z'' \qquad\qquad I_2\varphi_2'' = z''$$

$$\qquad\qquad\qquad\qquad\qquad\qquad\qquad\qquad\qquad\qquad (2.4\text{-}50)$$

$$I_2\omega_2 = I_1\omega_1 \qquad\qquad I_2\omega_2' = I_1\omega_1' \qquad\qquad I_2\varphi_2'' = I_1\varphi_1''$$

Using the above equalities and eliminating the expressions for F and S from the preceding three equations, gives the final model for the load's height change in the form of the following second-order OLDE

$$\underbrace{(M + \frac{J_1}{I_1^2} + \frac{J_2}{I_2^2})I_1 z''}_{a_1} = M_z - Mg\,I_1 \quad, \tag{2.4-51}$$

or the state-space equation

$$\begin{bmatrix} z \\ w \end{bmatrix}' = \underbrace{\begin{bmatrix} 0 & 1 \\ 0 & 0 \end{bmatrix}}_{\mathbf{A}} \begin{bmatrix} z \\ w \end{bmatrix} + \begin{bmatrix} 0 & 0 \\ \underbrace{\frac{1}{a_1}}_{b_{21}} & \underbrace{-\frac{g I_1}{a_1}}_{b_{22}} \end{bmatrix} \begin{bmatrix} M_z \\ M \end{bmatrix} \tag{2.4-52}$$

The eigenvalues of matrix $\mathbf{A}$ are equal

$$\lambda_1 = \lambda_2 = 0 \quad . \tag{2.4-53}$$

Thus we have a series of two "pure" integral systems with no time lag. These interrelations are clarified in the block diagram presentation in Figure 2.4-12.

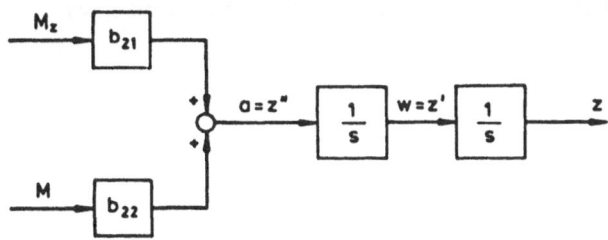

Fig. 2.4-12 Block diagram of load-lifting system

Figure 2.4-12 shows quite clearly that a load of mass M will rest only if

$$M_z b_{21} + M b_{22} = 0$$

i.e. for

$$M_z = Mg \, r_1 \qquad\qquad\qquad\qquad (2.4\text{-}54)$$

In all other cases the load raises or descends continuously to (theoretic) infinity, which is a typical and generally known property of integral systems.

c) The forces caused by motion of this mechanical system arise in the damper and in the spring. According to (2.4-17), the force in the damper is proportional to the velocity difference between the piston and the cylinder

$$F_p = D(z' - L\varphi') \qquad\qquad\qquad\qquad (2.4\text{-}55)$$

The force in the spring is proportional to its compression

$$F_o = cL\varphi \quad . \qquad\qquad\qquad\qquad (2.4\text{-}56)$$

The equality of those two forces gives

$$D(z' - L\varphi') = cL\varphi \quad . \qquad\qquad\qquad\qquad (2.4\text{-}57)$$

i.e.

$$\frac{D}{c}\varphi' + \varphi = \frac{D}{cL}z' \qquad\qquad\qquad\qquad (2.4\text{-}58)$$

If we use the first column of Table 2.4-2 and introduce dynamic coefficients of inertia and resistance (2.4-58), becomes

$$\frac{I}{R}\varphi' + \varphi = \frac{1}{L}\frac{I}{R}z' \qquad\qquad\qquad\qquad (2.4\text{-}59)$$

i.e.

$$T\frac{d\varphi}{dt} + \varphi = \frac{1}{L}Tz' \quad . \qquad\qquad\qquad\qquad (2.4\text{-}60)$$

The time constant $T = I/R$ has already been obtained (see (2.2-1.24)), and the right-hand side of Equation (2.4-60) shows that the relationship between the displacement of cylinder piston z and angle of displacement φ is given by a derivative system with first-order lag. In the dynamic sense this relationship is completely analogous to that existing between outflow m_o and pressure P_o or the cross-sectional area of valve A_o, given by Equation (2.2-2.30) in the case of the gas storage tank. There is also a dynamic analogy with the process of heat conduction, with the relationship between heat flow rate q_2 and temperature of heated fluid ϑ_f, which is shown with an approximated transfer function (3.2-51), i.e. (3.2-61).

Eigenvalue $\lambda_1 = - 1/T$, but the dynamics of this process is also characterized by the zero value of the transfer function nominator, which is now in the origin, $n_1 = 0$.

We must also say that dynamic processes in mechanical systems will always be of a derivative character when the input variable is the displacement of the damper piston.

========

The next and last example in this section shows a case that in fact belongs to the following part of this chapter about processes with lumped parameters, the part in which composed processes of higher order will be shown. This example can serve as a kind of transition to the description of such processes, which can be broken down into several basic ones and whose model is (at least) of second or higher order. Besides, we also want to show an example here of a process that is originally nonlinear and has two states of equilibrium. One of those equilibrium points of operation is unstable, and the other marginally stable in this idealized frictionless case, but stable in practice. We also wish to show that the same models for the motion of composed mechanical systems can be obtained if we use Lagrange's equations of motion or Newton's laws. Although the example will be rather idealized and seemingly only of theoretic importance, we must say that these are cases that appear as problems of control in modern technical practice.

========

Example 4 Dynamics of a composed mechanical system. Stability of moving pendulum equilibrium state

Figure 2.4-13 shows an idealized pendulum whose mass is concentrated at its end, attached to a cart that can move in the vertical z-y plane under the influence of driving force F in the direction of the horizontal z axis. Two basic positions are shown in which the pendulum must be maintained by changes in force F. (The first, upright position represents the simplest one-dimensional model that is used for controlling rockets during takeoff, when they must be maintained in an upright position. The second is the case of a suspended pendulum, today often used as a model for planning control procedures when loads are transported by crane and the task of their optimum control can be regarded either as a question of the smallest possible swinging of the load, or as controlling load transport in minimum time.) Losses due to friction in the joints and the body's resistance to motion are neglected.

Derive the nonlinear model of motion dynamics for the moving pendulum, linearize it, formulate state-space equations, determine eigenvalues of the system matrix **A** and on their basis conclude about the characteristics of the transient process and the stability of the equilibrium point. Derive the models:

a - using Newton's laws (upright pendulum)
b - formulating Lagrange's equations of motion (suspended pendulum)

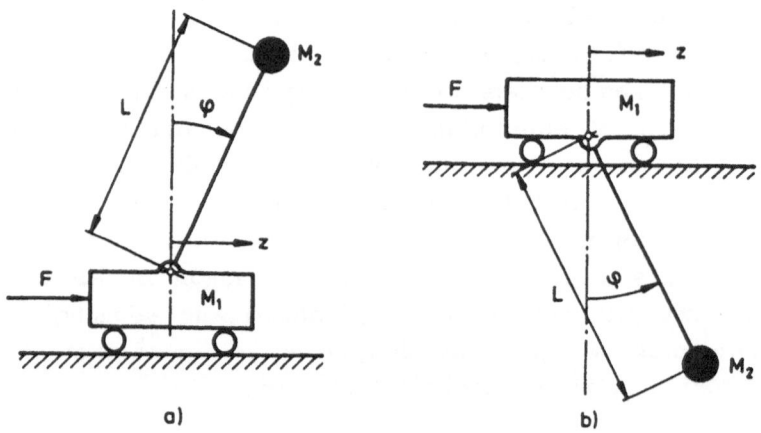

Fig. 2.4-13 Moving pendulum

a) Figure 2.4-14 shows the geometrical relations which will be used to derive the laws of motion. We must immediately observe that if the cart moves along the path z, the displacement of the pendulum in the direction of the z axis equals $z + L \sin \varphi$.

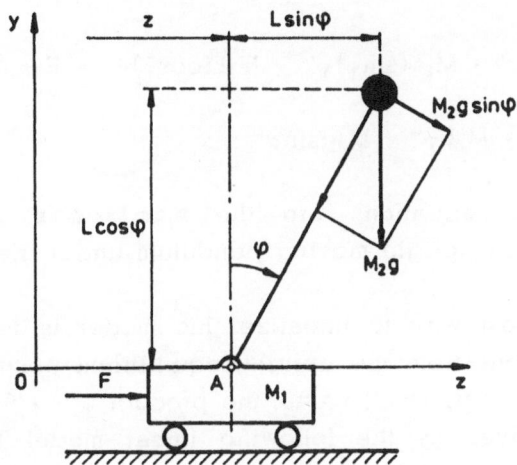

Fig. 2.4-14 Geometrical relations in an upright moving pendulum

The application of Newton's laws of motion for translation in the direction of the z axis yields

$$M_1 \frac{d^2z}{dt^2} + M_2 \frac{d^2}{dt^2} (z + L\sin\varphi) = F \quad . \tag{2.4-61}$$

For rotation of the pendulum about point A the law for the conservation of momentum is

$$\left[M_2 \frac{d^2}{dt^2}(z + L\sin\varphi) \right] L\cos\varphi - \left[M_2 \frac{d^2}{dt^2}(L\cos\varphi) \right] L\sin\varphi = M_2 g L\sin\varphi \tag{2.4-62}$$

Before differentiating the above expressions we must mention that the pendulum's angular displacement is a time function (a dynamic variable), which means

$$\frac{d}{dt}\sin\varphi = \cos\varphi \cdot \varphi'$$

$$\frac{d^2}{dt^2}\sin\varphi = -\sin\varphi \cdot \varphi'^2 + \cos\varphi \, \varphi'' \tag{2.4-63}$$

$$\frac{d}{dt}\cos\varphi = -\sin\varphi \cdot \varphi'$$

$$\frac{d^2}{dt^2}\cos\varphi = -\cos\varphi \cdot \varphi'^2 - \sin\varphi \cdot \varphi'' \quad .$$

Referring to the above expressions and differentiating the equations yields

$$(M_1 + M_2)z'' - M_2L(\sin\varphi)\varphi'^2 + M_2L(\cos\varphi)\varphi'' = F \qquad (2.4\text{-}64)$$

$$M_2z''\cos\varphi + M_2L\varphi'' = M_2g\sin\varphi \quad . \qquad (2.4\text{-}65)$$

These two equations are the **nonlinear** model for the motion dynamics of an upright moving pendulum under the influence of force F.

The simplest way to linearize this model is to use the fact that for small deviations from the upright, equilibrium position $\sin\varphi = \varphi$, $\cos\varphi = 1$, and because of its small value, the product $\varphi \cdot \varphi'^2$ can be made equal to zero. This gives us the following linear model for the motion of this system, which is satisfactory for small angular displacements.

$$(M_1 + M_2)z'' + M_2L\varphi'' = F \quad . \qquad (2.4\text{-}66)$$

$$M_2z'' + M_2L\varphi'' = M_2g\varphi \qquad (2.4\text{-}67)$$

To obtain models in the form of matrix state-space equations the state variables must be selected. As until now, in mechanical processes of motion the most natural selection will also be made. Here state variables are the displacements and velocities of the cart and the pendulum. Thus for

$$
\begin{aligned}
x_1 &= z \\
x_2 &= z' = w \\
x_3 &= \varphi \\
x_4 &= \varphi' = \omega
\end{aligned}
\qquad (2.4\text{-}68)
$$

the linear model obtained can be transformed into this final form

$$
\begin{bmatrix} z \\ w \\ \varphi \\ \omega \end{bmatrix}'
=
\begin{bmatrix}
0 & 1 & 0 & 0 \\
0 & 0 & -\dfrac{M_2}{M_1}g & 0 \\
0 & 0 & 0 & 1 \\
0 & 0 & \dfrac{M_1+M_2}{M_1L}g & 0
\end{bmatrix}
\cdot
\begin{bmatrix} z \\ w \\ \varphi \\ \omega \end{bmatrix}
+
\begin{bmatrix} 0 \\ \dfrac{1}{M_1} \\ 0 \\ -\dfrac{1}{M_1L} \end{bmatrix}
\begin{bmatrix} F \end{bmatrix}
\qquad (2.4\text{-}69)
$$

X' = A X + B U

Solving equation $(\lambda I - A) = 0$ gives the following characteristic equation for that model

$$\lambda^2(\lambda^2 - \frac{M_1+M_2}{M_1 L}g) = 0 \quad . \tag{2.4-70}$$

from which the following eigenvalues follow directly

$$\lambda_1 = 0$$
$$\lambda_2 = 0$$

$$\lambda_3 = \sqrt{\frac{M_1+M_2}{M_1 L}g}$$

$$\tag{2.4-71}$$

$$\lambda_4 = -\sqrt{\frac{M_1+M_2}{M_1 L}g} = -\lambda_3$$

As until now in the case of translation, the eigenvalues $\lambda_1 = \lambda_2 = 0$ belong to changes in the position z and velocity w of the cart. The eigenvalue λ_3 **is positive**, and shows that the **equilibrium position of the upright pendulum is unstable** in the sense that if one variable (in this case the angular displacement of pendulum φ) is disturbed for even the smallest amount it will irreversibly depart further and further away from the equilibrium position $\varphi = 0$. λ_4 is a real and negative variable and related to changes in angular velocity.

b) The Lagrange equations of motion are given as follows

$$\frac{d}{dt}(\frac{\partial L}{\partial q_i'}) - \frac{\partial L}{\partial q_i} + \frac{\partial D}{\partial q_i'} = Q_i , (i = 1, 2, \ldots , k) \tag{2.4-72}$$

L ... Lagrangian, L = T - V
T ... kinetic energy
V ... potential energy
D ... Rayleigh's dissipation function
Q ... external (generalized) force in the direction of the i-th coordinate
q ... generalized coordinate
k ... degree of freedom of the mechanical system

Equation (2.4-72) shows a system of k (in the general case nonlinear) differential equations of the second order.

The moving suspended pendulum is an example of a mechanical system with two degrees of freedom. The generalized coordinates of that motion are variable z for the translation and the angular displacement φ for the rotation of the pendulum.

The kinetic energy of the system is

$$T = \frac{1}{2}M_1 w_1^2 + \frac{1}{2}M_2 w_2^2 \quad . \tag{2.4-73}$$

The velocity of the cart $w_1 = z$, and w_2 is the absolute velocity of mass M_2 which is obviously composed of velocity components in the direction of the z axis and the y axis. Therefore, Figure 2.4-14 and Equation (2.4-63) yield

$$w_2^2 = (z + L\sin\varphi)'^2 + (L\cos\varphi)'^2 = z'^2 + L^2\varphi'^2 + 2z'L\cos\varphi \cdot \varphi' \tag{2.4-74}$$

Substituting this expression for velocity into (2.4-73) yields

$$T = \frac{1}{2}M_1 w_1^2 + \frac{1}{2}M_2(z'^2 + L^2\varphi'^2 + 2z'L\cos\varphi \cdot \varphi') \tag{2.4-75}$$

The potential energy of the system is

$$V = M_2 gL(1 - \cos\varphi) \quad . \tag{2.4-76}$$

It is clear that in the equilibrium state $V = 0$, which results from the upper equation for $\varphi = 0$.

The Lagrangian is now not difficult to calculate

$$L = T - V = \frac{1}{2}M_1 w_1^2 + \frac{1}{2}M_2(z'^2 + L^2\varphi'^2 + 2z'L\cos\varphi \cdot \varphi') - M_2 gL(1 - \cos\varphi)$$

$$\tag{2.4-77}$$

The equations of motion along the generalized coordinates follow directly from (2.4-72)

$$\frac{d}{dt}\left(\frac{\partial L}{\partial z'}\right) - \frac{\partial L}{\partial z} = F \ ,$$

$$\frac{d}{dt}\left(\frac{\partial L}{\partial \varphi'}\right) - \frac{\partial L}{\partial \varphi} = 0 \quad . \tag{2.4-78}$$

Substituting Equation (2.4-77) for L into this system of equations and carrying out the necessary operations gives the demanded description of the motion dynamics of a moving suspended pendulum

$$(M_1 + M_2)z'' - M_2L(\sin\varphi)\varphi'^2 + M_2L(\cos\varphi)\varphi'' = F \quad , \qquad (2.4\text{-}79)$$

$$M_2z''\cos\varphi + M_2L\varphi'' = - M_2g\sin\varphi \quad . \qquad (2.4\text{-}80)$$

Again this is a model composed of two nonlinear second-order ODE, which differs from the model of the upright pendulum in the sign of the right-hand side of the second equation. However, this "small" difference leads to essentially different results in the way in which this system moves.

Linearization and the selection of state variables is carried out in the same way as in the preceding case, which gives the following linear model for a moving suspended pendulum in the form of matrix state-space equations

$$
\begin{bmatrix} z \\ w \\ \varphi \\ \omega \end{bmatrix}' =
\begin{bmatrix}
0 & 1 & 0 & 0 \\
0 & 0 & \dfrac{M_2}{M_1}g & 0 \\
0 & 0 & 0 & 1 \\
0 & 0 & -\dfrac{M_1+M_2}{M_1L}g & 0
\end{bmatrix}
\cdot
\begin{bmatrix} z \\ w \\ \varphi \\ \omega \end{bmatrix}
+
\begin{bmatrix} 0 \\ \dfrac{1}{M_1} \\ 0 \\ -\dfrac{1}{M_1L} \end{bmatrix}
\begin{bmatrix} F \end{bmatrix}
\qquad (2.4\text{-}81)
$$

The characteristic equation of matrix **A** is now

$$\lambda^2(\lambda^2 - a_{43}) = 0 \quad . \qquad (2.4\text{-}82)$$

from which follow the eigenvalues for a moving suspended pendulum

$$\lambda_1 = 0$$
$$\lambda_2 = 0$$
$$\lambda_3 = \sqrt{\frac{M_1+M_2}{M_1L}g} \quad j$$

$$\qquad (2.4\text{-}83)$$

$$\lambda_4 = -\sqrt{\frac{M_1+M_2}{M_1L}g} \quad j \quad .$$

The eigenvalues λ_1 and λ_2 belong to translational motion of the cart and have not changed, which was to be expected since there was no change in that motion.

However, the change in the second pair of eigenvalues is essential. λ_3 and λ_4 are a complex conjugate pair on the imaginary axis (their real part equals zero), which means they characterize an undamped periodical transient process similar to processes that have already been encountered.

The frequency ω of the oscillation (swinging) is

$$\omega = \sqrt{\frac{M_1 + M_2}{M_1 L} g}$$.   (2.4-84)

These eigenvalues, neither of which is in the right-hand semiplane of the complex plane of characteristic Equation (2.4-82) solutions, show that the position of equilibrium for the suspended pendulum ($\varphi = 0$) is no longer unstable. All the same, since all possible resistances have been neglected, we must say that this position is not asymptotically stable in the sense that after a disturbance the angular displacement of the pendulum will gradually return to its initial value $\varphi = 0$. With this choice of assumptions, after disturbance the pendulum will swing with frequency ω around the initial position of equilibrium without damping.

B. COMPLEX SYSTEMS OF HIGHER ORDER

2.5 EXAMPLES OF COMPLEX SYSTEMS

In all the preceding sections of this chapter on processes with lumped parameters (with the exception of the last example in Section 2.4, which represents a transition to descriptions of higher-order processes) we met with processes whose dynamics was described by DE of no more than second order. Such processes have been named basic, which allows us to conclude that they could be parts of even the most complex systems, i.e. that complex systems can be reduced to various combinations of those basic processes. This is very frequently done in practice where models of extremely complicated technological systems are encountered. For certain purposes of analysis or synthesis of control algorithms such systems are described by models of fourth, third, as well as second or first order. Depending on the demands and goals that must be satisfied, opposite situations also exist when a seemingly simple object is described by a dynamic model of high order. The concepts of simple and complex are obviously relative and we must distinguish, especially from the aspect of simplicity, the process from its model. All combinations are possible: complex process - simple model, simple process - complex model, and so on. This section will, therefore, analyze cases when the model is of second or of more than second order, and the procedure for obtaining models and the usual analysis will be shown on several simple cases.

Complex processes described by higher-order models are met in various situations. In modeling the dynamics of a large plant it is usual to break its structure down into elementary devices: pumps, reactors, pipes, valves and the like, and then model each of those parts separately. Interconnecting all the basic models gives a high-order model for the whole plant. Another case when a high-order model is obtained is for processes that occur in objects of simple geometrical structure within which simultaneous mass, energy and momentum changes take place. Deriving all the demanded equations leads to a higher-order model. Finally, in technical practice the dynamics of processes with distributed parameters is usually investigated using spatial discretization. This means that a continuous process is divided into elements of finite dimension and then balance (conservation) equations are derived for each of those elements. There is also another way: first PDE are formulated describing the dynamics of the distributed process and then mathematical discretization is performed with respect to the spatial variables. Both procedures lead to the same result in the sense that instead of one or several PDE a system of ODE is obtained, which can then be elaborated further analytically or by numerical processing on a computer.

Thanks to calculating aids - computers, today models of very high order are derived and processed. However, for understanding the basic properties of a model it is irrelevant whether it is of 5th, 55th, 105th or 1005th order. With very high orders new problems connected with computer processing are encountered, but these are not the concern of this book. Therefore, the following examples will remain simple so as to be clear and easy to understand. They certainly include the example of the moving pendulum from the preceding section.

===============================

Example 1 Three liquid tanks in series

Derive a mathematical model of dynamics for three liquid tanks in series, Figure 2.5-1, and after linearization determine system matrix **A** input matrix **B** and find the eigenvalues. Neglect liquid inertia in tanks and pipes. The following data are given:

$$A_1 = 1 \text{ m}^2 \qquad d_1 = 0.045 \text{ m} \qquad L_1 = 40 \text{ m} \qquad \lambda_1 = 0.04 \qquad m_{i1} = 3 \frac{\text{kg}}{\text{s}}$$

$$A_2 = 0.5 \text{ m}^2 \qquad d_2 = 0.05 \text{ m} \qquad L_2 = 1 \text{ m} \qquad \lambda_2 = 0.04 \qquad m_{i2} = 2 \frac{\text{kg}}{\text{s}}$$

$$A_3 = 1 \text{ m}^2 \qquad d_3 = 0.06 \text{ m} \qquad L_3 = 20 \text{ m} \qquad \lambda_3 = 0.04 \qquad m_{i3} = 2 \frac{\text{kg}}{\text{s}}$$

The pressure drop on the valve is $\delta P_v = 1$ bar and the mean cross-sectional area of the valve is $a_v = 0.001 \text{ m}^2$.

The dynamic model is obtained in the usual way by deriving equations for the conservation of mass for each of the tanks. Besides those equations, however, it is also necessary to derive algebraic expressions for liquid flow through the system using Bernoulli's equations.

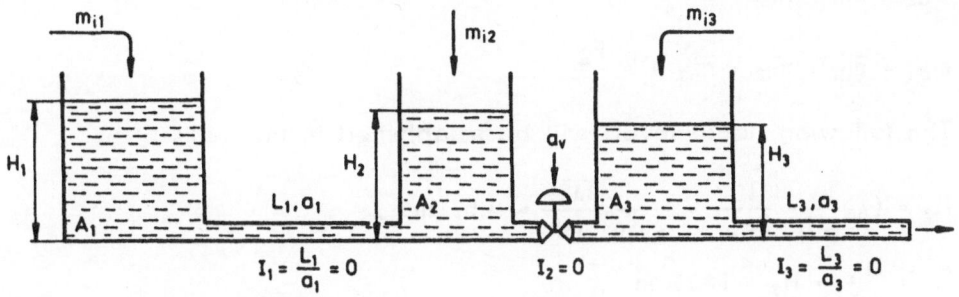

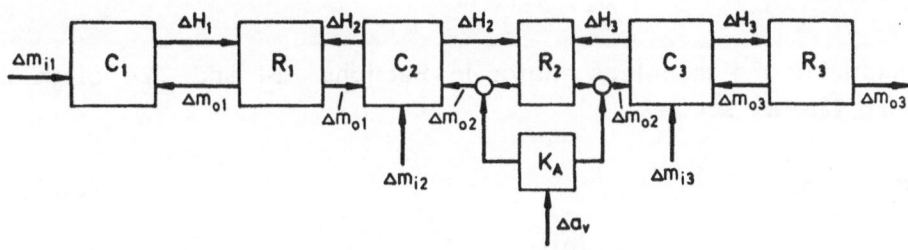

Fig. 2.5-1 Three liquid tanks in series and a block diagram
of the interrelations

$$m_{i1} - m_{o1} = A_1\rho\frac{dH_1}{dt} \quad . \tag{2.5-1}$$

$$m_{i2} + m_{o1} - m_{o2} = A_2\rho\frac{dH_2}{dt} \quad . \tag{2.5-2}$$

$$m_{i3} + m_{o2} - m_{o3} = A_3\rho\frac{dH_3}{dt} \quad . \tag{2.5-3}$$

$$\rho g(H_1 - H_2) = \delta P_{c1} = \lambda_1\frac{L_1}{d_1}\frac{1}{2a_1^2\rho}m_{o1}^2 \quad . \tag{2.5-4}$$

$$\rho g(H_2 - H_3) = \delta P_v \quad . \tag{2.5-5}$$

$$\rho gH_3 = \frac{w_3^2}{2}\rho + \delta P_{c3} = \frac{m_{o3}^2}{2a_3^2\rho} + \lambda_3\frac{L_3}{d_3}\frac{1}{2a_3^2\rho}m_{o3}^2 \quad . \tag{2.5-6}$$

The steady state for the given mass flow rate values and the tank system must first be determined, and then linearization about that operating point performed. Equalizing the right-hand sides of the first three equations with zero yields

$$\bar{m}_{i1} = \bar{m}_{o1} = 3 \frac{kg}{s} \quad ,$$

$$\bar{m}_{o2} = \bar{m}_{i1} + \bar{m}_{i2} = 5 \frac{kg}{s} \quad ,$$

$$\bar{m}_{o3} = \bar{m}_{i1} + \bar{m}_{i2} + \bar{m}_{i3} = 7 \frac{kg}{s} \quad .$$

The following liquid levels will be established in the tanks

$$H_3 = (\lambda_3 \frac{L_3}{d_3} \frac{1}{2a_3^2 \rho} + \frac{1}{2a_3^2 \rho}) \frac{\bar{m}_{o3}^2}{\rho g} = 4.57 \text{ m} \quad ,$$

$$H_2 = \frac{\delta P_v}{\rho g} + H_3 = 14.77 \text{ m} \quad ,$$

$$H_1 = \lambda_1 \frac{L_1}{d_1} \frac{1}{2a_1^2 \rho} \frac{\bar{m}_{o1}^2}{\rho g} + H_2 = 21.21 \text{ m} \quad .$$

Linearizing the model as shown in Sections 2.1-1 and 2.2-1 gives the following **linear model**

$$\Delta m_{i1} - \Delta m_{o1} = C_1 \frac{d\Delta H_1}{dt} \tag{2.5-7}$$

$$\Delta m_{i2} + \Delta m_{o1} - \Delta m_{o2} = C_2 \frac{d\Delta H_2}{dt} \tag{2.5-8}$$

$$\Delta m_{i3} + \Delta m_{o2} - \Delta m_{o3} = C_3 \frac{d\Delta H_3}{dt} \tag{2.5-9}$$

$$\Delta H_1 - \Delta H_2 = R_1 \Delta m_{o1} \tag{2.5-10}$$

$$\Delta H_2 - \Delta H_3 = R_2 \Delta m_{o2} - R_2 K_A \Delta a_v \tag{2.5-11}$$

$$\Delta H_3 = R_3 \Delta m_{o3} \tag{2.5-12}$$

The resistance and capacitance are obtained from the known expressions

$$R_1 = \frac{2(H_1 - H_2)}{\bar{m}_{o1}} = 4.3 \qquad C_1 = A_1 \rho = 1000 \qquad T_1 = 4300$$

$$R_2 = \frac{2(H_2 - H_3)}{\bar{m}_{o2}} = 4.08 \qquad C_2 = 500 \qquad T_2 = 2040$$

$$R_3 = \frac{2H_3}{\bar{m}_{o3}} = 1.3 \qquad C_3 = 1000 \qquad T_3 = 1300$$

$$K_A = \frac{\bar{m}_{o2}}{a_v} = 5000$$

If the internal variables m_{o1}, m_{o2} and m_{o3} are excluded from the last system of equations, a little rearrangement yields this state-space description

$$
\begin{bmatrix} \Delta H_1 \\ \Delta H_2 \\ \Delta H_3 \end{bmatrix}' = \begin{bmatrix} -\dfrac{1}{T_1} & \dfrac{1}{T_2} & 0 \\[2ex] \dfrac{\frac{R_2}{R_1}}{T_2} & \dfrac{-(\frac{R_2}{R_1}+1)}{T_2} & \dfrac{1}{T_2} \\[2ex] 0 & \dfrac{\frac{R_3}{R_2}}{T_3} & \dfrac{-(\frac{R_3}{R_2}+1)}{T_3} \end{bmatrix} \begin{bmatrix} \Delta H_1 \\ \Delta H_2 \\ \Delta H_3 \end{bmatrix} +
$$

$$
\qquad\qquad\qquad\qquad\qquad \mathbf{A}
$$

$$
+ \begin{bmatrix} \dfrac{1}{C_1} & 0 & 0 & 0 \\[2ex] 0 & \dfrac{1}{C_2} & 0 & -\dfrac{K_A}{C_2} \\[2ex] 0 & 0 & \dfrac{1}{C_3} & \dfrac{K_A}{C_3} \end{bmatrix} \begin{bmatrix} \Delta m_{i1} \\ \Delta m_{i2} \\ \Delta m_{i3} \\ \Delta a_v \end{bmatrix} \qquad\qquad (2.5\text{-}13)
$$

After substituting into matrix $\mathbf{A}$ the values of the coefficients, a computer can be used to obtain the numerical values of the eigenvalues for this case

$$\lambda_1 = -0.000086, \quad \lambda_2 = -0.000736, \quad \lambda_3 = -0.001378 \ .$$

Three negative, real eigenvalues show that this is a stable equilibrium state of three first-order proportional systems in series, where there is feedback action of each successive member on the preceding one. This can be seen on the block diagram on Figure 2.5-1.

=====

Example 2 Two-phase fluid tank

Figure 2.5-2 shows a two-phase fluid tank with completely separated phases. It is fed by m_{i1} kg/s saturated water and m_{i2} water and steam mixture with steam quality x and enthalpy h_2. m_{o1} kg/s dry saturated steam and m_{o2} kg/s saturated water are taken from the tank. Devices that are not shown on Figure 2.5-2 completely separate the mixture m_{i2} into the liquid and the vapor phase. Derive the model for unsteady changes of pressure P and the liquid phase level H, i.e. the water volume V_w. All the thermodynamic state equations for saturation variables are known, i.e. there are analytic expressions of the type $h' = h'(P)$, $\rho' = \rho'(P)$, $v'' = v''(P)$ and so on. The basic assumption, besides the mentioned phase separation, is that there is no mass transfer on the contact surface due to pressure change, and that both phases are homogeneous, i.e. there are no water drops in the steam, nor can steam bubbles form in the water.

The model will be derived by the same method that was used in Example 1, Section 2.1-2, when a dynamic model was obtained for a gas storage tank. In that case, however, because isothermal processes in the tank were assumed, no energy equation was formulated. Since temperature dynamics was neglected in Equation (2.1-2.4) the model for pressure dynamics was of first order and obtained only from the law for the conservation of mass and the gas equation. Here the situation is somewhat more complex.

The total mass M of the fluid and its internal energy U, i.e. the heat within the tank of constant volume V, are completely determined by two variables: pressure P and the volume of liquid phase V_w, M = $g_1(P,V_w)$, U = $g_2(P,V_w)$. In conditions of mass and energy flow equilibrium these are constant values. If the flows become unbalanced unsteady phenomena occur and to describe them **equations for the conservation of mass and energy** must be formulated. For the tank under consideration here they are

$$m_{i1} + m_{i2} - m_{o1} - m_{o2} = \frac{dM}{dt} \quad , \tag{2.5-14}$$

$$m_{i2}h_2 - (m_{o2} - m_{i1})h' - m_{o1}h'' = \frac{dU}{dt} \tag{2.5-15}$$

The total mass and energy in the tank comprises the mass and energy of the liquid and the vapor phase

$$M = V_w\rho' + (V - V_w)\rho'' \quad , \tag{2.5-16}$$

$$U = V_w\rho'u' + (V - V_w)\rho''u'' \quad . \tag{2.5-17}$$

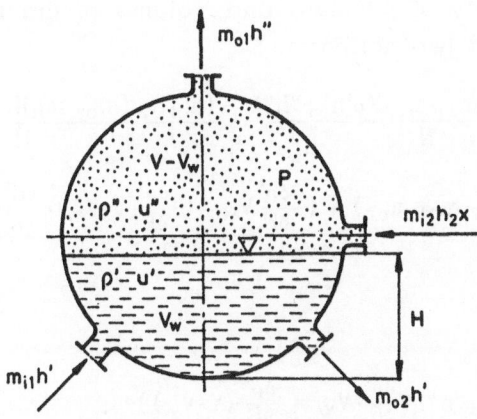

Fig. 2.5-2 Two-phase fluid tank

Substituting the known equality

$$\rho u = \rho h - P \quad ,$$
(2.5-18)

into the preceding expression, we get

$$U = V_w \rho'h' + (V - V_w)\rho''h'' - VP \quad .$$
(2.5-19)

Here we must point out that in literature, internal energy is often substituted by enthalpy, which would (had it been done in this derivation) lead to the disappearance of the last term on the right-hand side of Equation (2.5-19). This substitution leads to the appearance of a "dynamic" error in the "dynamic" term of Equation (2.5-15) (the term on its right-hand side) thus influencing the transient process but not the steady state as well.

For the needs of further derivation we must remember that the state variables on the boundary curves are functions of pressure only, and for all of them we can write

$$\frac{d}{dt} = \frac{\partial}{\partial P} \frac{dP}{dt}$$
(2.5-20)

Substituting (2.5-16) and (2.5-19) into the first two equations of this example, referring to (2.5-20), after a little rearrangement we get the model for the dynamics of pressure and volume of the liquid phase in the form of the following two nonlinear DE

$$\frac{dP}{dt} = \frac{(m_{i1}+m_{i2}-m_{o1}-m_{o2})(\rho'h'-\rho''h'')}{N} - \frac{(m_{i2}h_2-(m_{o2}-m_{i1})h'-m_{o1}h'')(\rho'-\rho'')}{N} \tag{2.5-21}$$

$$\frac{dV_w}{dt} = \frac{(m_{i1}+m_{i2}-m_{o1}-m_{o2}) - (\frac{\partial\rho'}{\partial P}V_w + \frac{\partial\rho''}{\partial P}(V-V_w))\frac{dP}{dt}}{\rho'-\rho''} \tag{2.5-22}$$

$$N = N_1 - N_2 \tag{2.5-23}$$

$$N_1 = (\rho'h' - \rho''h'')(\frac{\partial\rho'}{\partial P}V_w + \frac{\partial\rho''}{\partial P}(V-V_w)) \quad , \tag{2.5-24}$$

$$N_2 = (\rho'-\rho'')\left[\frac{\partial\rho'}{\partial P}V_w h' + \frac{\partial h'}{\partial P}V_w\rho' + \frac{\partial\rho''}{\partial P}(V-V_w)h'' + \frac{\partial h''}{\partial P}(V-V_w)\rho'' - V\right] \tag{2.5-25}$$

Equations (2.5-21) and (2.5-22) are the demanded nonlinear mathematical model of dynamic processes occurring in the two-phase fluid tank. As the given expressions show, to solve them it is necessary to have all the functional dependencies of thermodynamic saturation variables and their derivatives with reference to pressure P. By their nature these are clumsy expressions, and even in the case of their simplest linear approximation the above system of DE is practically insolvable without a digital computer.

But in this case the steady state can also be calculated analytically. After the right-hand sides of (2.5-14) and (2.5-15) are made equal to zero, these obvious equalities result

$$\overline{m}_{i1} + \overline{m}_{i2} = \overline{m}_{o1} + \overline{m}_{o2} \quad , \tag{2.5-26}$$

$$\overline{m}_{i2}\overline{h}_2 = (\overline{m}_{o2} - \overline{m}_{i1})\overline{h}' + \overline{m}_{o1}\overline{h}'' \quad . \tag{2.5-27}$$

The last expressions are completely clear and represent the equilibrium of mass and energy flows in the given steady state. Here it is important to point out that a new steady state will establish itself only in the case of limited change in h_2. In the case of a finite change in any other input (disturbance) variable (m_{i1}, m_{i2}, m_{o1}, m_{o2}) no steady state can be established and the tank shows properties of an integral system. In such cases mass and energy equilibrium are permanently disturbed and the variable V_w converges to theoretically infinite change, which would in practice lead to the complete disappearance of the liquid or the vapor phase in the tank.

The same conclusion is reached after the **model is linerized.** The functional dependence of variables V_w and P given in the model can also be expressed as follows

$$\frac{dV_w}{dt} = f_1 (V_w, P, m_{i1}, m_{i2}, m_{o1}, m_{o2}, h_2) \qquad (2.5\text{-}28)$$

$$\frac{dP}{dt} = f_2 (\underbrace{V_w, P}_{\text{state}}, \underbrace{m_{i1}, m_{i2}, m_{o1}, m_{o2}, h_2}_{\text{input}}) \qquad (2.5\text{-}29)$$

variables

Linearization is carried out in the usual manner by differentiating the above equations (see Appendix), but here we will not write out the complete analytical expressions for each specific coefficient because of their awkwardness. It is, nevertheless, useful to show what matrices A and B are built of in this case of a two-phase fluid tank.

$$\begin{bmatrix} \Delta V_w \\ \Delta P \end{bmatrix}' = \underbrace{\begin{bmatrix} \dfrac{\partial \bar{f}_1}{\partial V_w} & \dfrac{\partial \bar{f}_1}{\partial P} \\ \dfrac{\partial \bar{f}_2}{\partial V_w} & \dfrac{\partial \bar{f}_2}{\partial P} \end{bmatrix}_O}_{A} \begin{bmatrix} \Delta V_w \\ \Delta P \end{bmatrix} +$$

$$+ \underbrace{\begin{bmatrix} \dfrac{\partial \bar{f}_1}{\partial m_{i1}} & \dfrac{\partial \bar{f}_1}{\partial m_{i2}} & \dfrac{\partial \bar{f}_1}{\partial m_{o1}} & \dfrac{\partial \bar{f}_1}{\partial m_{o2}} & \dfrac{\partial \bar{f}_1}{\partial h_2} \\ \dfrac{\partial \bar{f}_2}{\partial m_{i1}} & \dfrac{\partial \bar{f}_2}{\partial m_{i2}} & \dfrac{\partial \bar{f}_2}{\partial m_{o1}} & \dfrac{\partial \bar{f}_2}{\partial m_{o2}} & \dfrac{\partial \bar{f}_2}{\partial h_2} \end{bmatrix}_O}_{B} \begin{bmatrix} \Delta m_{i1} \\ \Delta m_{i2} \\ \Delta m_{o1} \\ \Delta m_{o2} \\ \Delta h_2 \end{bmatrix} \qquad (2.5\text{-}30)$$

The partial derivatives shown in Equation (2.5-30) are the differentiation of the right-hand side of Equation (2.5-22) for the first row of matrices A and B, and of the right-hand side of Equation (2.5-21) for the second row of these matrices, with respect to the variables of state and input.

The line above the functions and the subscript O show that the partial derivatives must be calculated in a specific operational (steady) state. It must also be said that the first column of matrix A is composed of zeros, i.e. the coefficients a_{11} and a_{21} are equal to zero, which shows that one of the matrix eigenvalues will be zero and to it will corespond an infinitely great time constant. In other words, the variable to which that eigenvalue is related needs an infinitely long time to reach a new steady state. Or to put it even more clearly - that variable cannot reach a new steady value in a finite period of time. That variable, as has already been said, is the volume of the liquid phase V_W. The second eigenvalue equals the coefficient a_{22} and in practice this is a negative real value characteristic of aperiodic proportional first- order systems. The process of pressure P changes in the tank will show these dynamic characteristics.

Finally, it must be repeated that since both mass and energy are stored in the tank, this is a second-order model for changes in the volume of liquid phase V_W (or in the liquid level H, because those two variables are uniquely linked by algebraic expressions) and pressure P in the tank. In this case, V_W shows properties of an integral and P of a proportional system.

Example 3 Gear train system

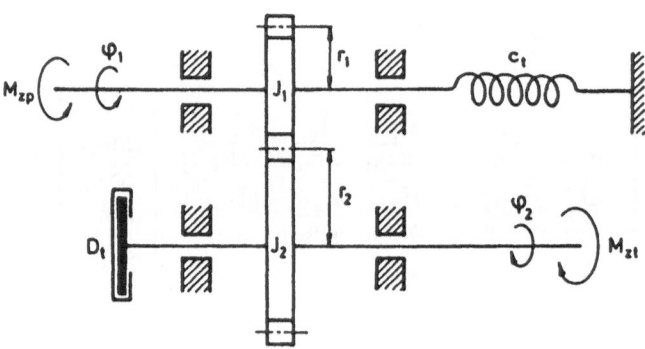

Fig. 2.5-3 Gear train system

Derive a model for dynamic changes of angular displacements and angular velocities for the gear train system shown on Figure 2.5-3, if the interrelations are linear. All the geometric variables and the coefficients c_t and D_t (torsional spring constant and torsional coefficient of viscous friction) are given. Loss due to the friction of shaft 1 and the stiffness of shaft 2 are neglected. The input disturbance variables are driving torque M_{zp} and load torque M_{zt}. The model is to be shown in the form of state-space equations.

To derive the dynamic model of this mechanical system we must formulate equations for momentum conservation for the rotation of shafts 1 and 2.

For shaft 1

$$J_1 \varphi_1'' = - c_t \varphi_1 - M_{z1} + M_{zp} \quad . \tag{2.5-31}$$

where M_{z1} is the torque transmitted to shaft 2, for which

$$J_2 \varphi_2'' = - D_t \varphi_2' - M_{zt} + M_{z2} \quad . \tag{2.5-32}$$

M_{z2} is transmitted to shaft 2 through gears.

There is also

$$I_1 \varphi_1 = I_2 \varphi_2 \quad . \tag{2.5-33}$$

$$M_{z1} \varphi_1' = M_{z2} \varphi_2' \quad . \tag{2.5-34}$$

The last equation expresses the equality of the work done by gear 1 and gear 2. The above equations yield

$$\frac{M_{z1}}{M_{z2}} = \frac{\varphi_2}{\varphi_1} = \frac{I_1}{I_2} \tag{2.5-35}$$

Referring to (2.5-35), we see that (2.5-32) becomes

$$J_2 \varphi_2'' + D_t \varphi_2' + M_{zt} = M_{z2} = M_{z1} \frac{I_2}{I_1} \tag{2.5-36}$$

If M_{z1} is expressed from (2.5-31) and substituted into the above expression, after substituting $\varphi_2 = (I_1/I_2) \varphi_1$, the DE describing the dynamics of the gear train system is obtained

$$\underbrace{(J_1 + (\frac{I_1}{I_2})^2 J_2)}_{J} \varphi_1'' + D_t (\frac{I_1}{I_2})^2 \varphi' + c_t \varphi_1 = M_{zp} - \frac{I_1}{I_2} M_{zt} \quad . \tag{2.5-37}$$

The last equation, with the equivalent substitution of φ_1 by φ_2 gives a model for the dynamics of the gear train system expressed by the angular displacement of the shaft φ_2

$$(J_1 + (\frac{r_1}{r_2})^2 J_2)\varphi_2'' + D_t(\frac{r_1}{r_2})^2 \varphi_2' + c_t \varphi_2 = \frac{r_1}{r_2} M_{zp} - (\frac{r_1}{r_2})^2 M_{zt} \qquad . \qquad (2.5\text{-}38)$$

It should be observed that the dynamic model of this gear train system is of the second order, but is expressed in two ways - through the angular displacement of shaft 1 and through the angular displacement of shaft 2. The model is thus **either** Equation (2.5-37) **or** Equation (2.5-38), and it would be wrong to consider that both equations at once represent the model demanded. In that case this would be a fourth-order system, which it is not. We can chose either the pair φ_1 and φ'_1 or the pair φ_2 and φ'_2 for the state variable. If we consider that the displacement angles φ_1 and φ_2 and the angular velocities φ'_1 and φ'_2 are output variables, the following two notations in the form of state-space equations are possible.

$$x_1 = \varphi_1 , \; x_2 = \varphi'_1 = \omega_1$$

State equation:

$$\begin{bmatrix} \varphi_1 \\ \\ \omega_1 \end{bmatrix}' = \begin{bmatrix} 0 & 1 \\ \\ -\frac{c_t}{J} & \frac{-D_t(\frac{r_1}{r_2})^2}{J} \end{bmatrix} \begin{bmatrix} \varphi_1 \\ \\ \omega_1 \end{bmatrix} + \begin{bmatrix} 0 & 0 \\ \\ \frac{1}{J} & -\frac{\frac{r_1}{r_2}}{J} \end{bmatrix} \begin{bmatrix} M_{zp} \\ \\ M_{zt} \end{bmatrix} \qquad (2.5\text{-}39)$$

Output equation:

$$\begin{bmatrix} \varphi_1 \\ \omega_1 \\ \varphi_2 \\ \omega_2 \end{bmatrix} = \begin{bmatrix} 1 & 0 \\ 0 & 1 \\ \frac{r_1}{r_2} & 0 \\ 0 & \frac{r_1}{r_2} \end{bmatrix} \begin{bmatrix} \varphi_1 \\ \omega_1 \end{bmatrix} \qquad (2.5\text{-}40)$$

$$x_1 = \varphi_2, \; x_2 = \varphi_2' = \omega_2$$

State equation:

$$
\begin{bmatrix} \varphi_2 \\ \omega_2 \end{bmatrix}' = \begin{bmatrix} 0 & 1 \\ -\dfrac{c_t}{J} & \dfrac{-D_t(\frac{r_1}{r_2})^2}{J} \end{bmatrix} \begin{bmatrix} \varphi_2 \\ \omega_2 \end{bmatrix} + \begin{bmatrix} 0 & 0 \\ \dfrac{r_1}{r_2} & \dfrac{-(\frac{r_1}{r_2})^2}{J} \end{bmatrix} \begin{bmatrix} M_{zp} \\ M_{zt} \end{bmatrix} \qquad (2.5\text{-}41)
$$

Output equation:

$$
\begin{bmatrix} \varphi_1 \\ \omega_1 \\ \varphi_2 \\ \omega_2 \end{bmatrix} = \begin{bmatrix} \dfrac{r_2}{r_1} & 0 \\ 0 & \dfrac{r_2}{r_1} \\ 1 & 0 \\ 0 & 1 \end{bmatrix} \begin{bmatrix} \varphi_2 \\ \omega_2 \end{bmatrix} \qquad (2.5\text{-}42)
$$

Both these selections of state variables are accurate and this example is useful to show how the selections can often be made in several ways. There is more on this subject in the Appendix, and here we must emphasize that in this case matrix **A** has remained unchanged, which is not always so. However, the basic properties of the process shown by the eigenvalues are always preserved. Here, in matrix **A** of Equations (2.5-39) and (2.5-41), they have obviously remained the same.

The following example will show how a high-order model can be obtained by the discretization of a spatially distributed process whose model is of infinite order. This infinite order results from the fact that every (even the simplest) distributed process is described by a partial DE showing the change of a certain variable in every point in the space in which that process occurs, and there is an unlimited number of such points. The order of the processes themselves, for the simplest PDE, is no problem when an analytic solution can be obtained. An example of how such a solution is obtained in a closed form is given in the following chapter, but it must be said that such cases are rare. Usually this is not the way to obtain a solution and the use of computers is unavoidable. The first step is in most cases to reduce the model, i.e. to decrease its order. Discretization with respect to the spatial variable is the usual way to transform one or several PDE into a finite system of ODE.

In the following example a hyperbolic first-order PDE (3.1-28) from the first section of the following chapter has been selected for discretization. It describes the dynamics of convective heat transfer to a fluid flowing through an uninsulated pipe. This is a linear PDE, which is suitable because problems of discretization will not unnecessarily be complicated by problems of linearization. These two procedures are completely independent from the aspect of the order in which they are performed, and when we need to formulate a nonlinear spatially distributed process in the form of state space, this can be done either by first performing discretization and then the linearization of the system of ODE obtained, or by proceeding in the opposite order. After the PDE has been linearized, a system of ODE is obtained through discretization, which is then easily transformed into matrix state-space notation.

We must also say that **discretization** will be performed by the **method of backward difference** which is only one of the many ways in which it can be done.

==

Example 4 Discretization of distributed process

For purposes of numerical simulation a hyperbolic first-order PDE (3.1-28) must be mathematically discretized. The equation describes the dynamics of the condenser heat exchanger ($\vartheta_w \neq f(z)$, $\vartheta_w = \vartheta_w(t)$) developed and shown on Figure 3.1-5 in the form of a pipe. For discretization into four elements, show the reduced model in the form of state-space equations.

Also show how the same model can be obtained if the process is physically divided first and the conservation equations formulated for each of the thus obtained finite elements.

The discretization of the following hyperbolic PDE

$$\frac{\partial \vartheta}{\partial t} + w\frac{\partial \vartheta}{\partial z} + k\vartheta = k\vartheta_w \quad , (k = \frac{\alpha U}{A\rho c}) \quad , \tag{2.5-43}$$

will be carried out by the **method of backward difference** by which the gradient of a specific function with respect to the space variable z, **in the section n**, is substituted in the following manner

$$\left(\frac{\partial f}{\partial z}\right)_n = \frac{f_n - f_{n-1}}{\delta z} \quad .$$

(2.5-44)

f_n —— function value in the section n

f_{n-1} —— function value in the section n-1

δz —— distance between sections n-1 and n (or, length of element obtained through discretization)

Besides this method there are also many other methods (for example the methods of forward and central difference) and manners which are essentially no different. We will not enter into an analysis of the advantages and disadvantages of such procedures nor will any other methods be shown here.

Applying (2.5-44) to the given PDE, the following system of n first-order ODE is obtained

$$\frac{d\vartheta_n}{dt} + w\frac{\vartheta_n - \vartheta_{n-1}}{\delta z} + k\vartheta_n = k\vartheta_w \quad , \ n = 2, \ NE + 1 \quad .$$

(2.5-45)

NE is the number of elements obtained by discretization (see Figure 2.5-4). The system of four ODE in the case of the division of the heat exchanger into four elements follows

$$\frac{d\vartheta_2}{dt} + k_1\vartheta_2 = k_2\vartheta_1 + k\vartheta_w$$

$$\frac{d\vartheta_3}{dt} + k_1\vartheta_3 = k_2\vartheta_2 + k\vartheta_w$$

$$\frac{d\vartheta_4}{dt} + k_1\vartheta_4 = k_2\vartheta_3 + k\vartheta_w$$

(2.5-46)

$$\frac{d\vartheta_5}{dt} + k_1\vartheta_5 = k_2\vartheta_4 + k\vartheta_w$$

$$\left(k_1 = \frac{w}{\delta z} + k, \ k_2 = \frac{w}{\delta z}\right)$$

If variables w, k and ϑ_w depended on spatial coordinate z, they would also have subscript n in (2.5-45), or a corresponding subscript in the system of equations (2.5-46). This system is not difficult to show in the form of matrix state-space equations if the only temperature of interest, temperature ϑ_5, is selected as the output variable

$$
\begin{bmatrix} \vartheta_2 \\ \vartheta_3 \\ \vartheta_4 \\ \vartheta_5 \end{bmatrix}' = \begin{bmatrix} -k_1 & 0 & 0 & 0 \\ k_2 & -k_1 & 0 & 0 \\ 0 & k_2 & -k_1 & 0 \\ 0 & 0 & k_2 & -k_1 \end{bmatrix} \begin{bmatrix} \vartheta_2 \\ \vartheta_3 \\ \vartheta_4 \\ \vartheta_5 \end{bmatrix} + \begin{bmatrix} k_2 & k \\ 0 & k \\ 0 & k \\ 0 & k \end{bmatrix} \begin{bmatrix} \vartheta_1 \\ \vartheta_w \end{bmatrix} \qquad (2.5\text{-}47)
$$

X' = A X + B U

$$
\begin{bmatrix} \vartheta_5 \end{bmatrix} = \begin{bmatrix} 0 & 0 & 0 & 1 \end{bmatrix} \begin{bmatrix} \vartheta_2 \\ \vartheta_3 \\ \vartheta_4 \\ \vartheta_5 \end{bmatrix} + \begin{bmatrix} 0 & 0 \end{bmatrix} \begin{bmatrix} \vartheta_1 \\ \vartheta_w \end{bmatrix} \qquad (2.5\text{-}48)
$$

Y = C X + D U

This shows how discretization transforms an infinite-dimensional process whose dynamics is described by PDE into a model of finite dimensions. Intuitively it is clear that a system of ODE will describe a process, the more faithfully the higher the degree of discretization, i.e. the "thicker" the division. But in that case the model obtained is of higher order and more difficult to process, so it is up to the model builder to decide on its final size.

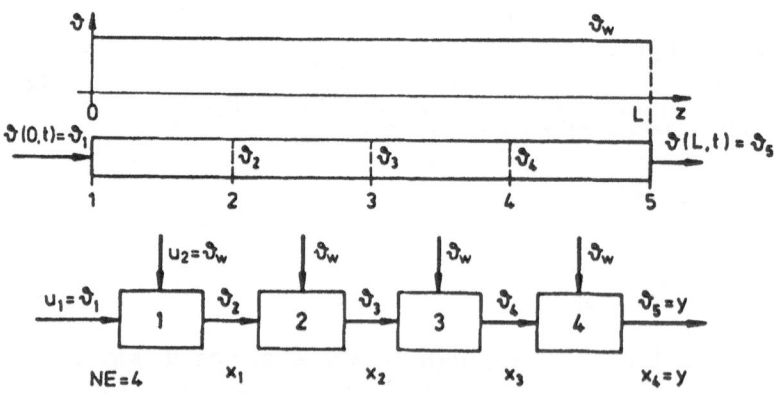

Fig. 2.5-4 Discretization of a spatially distributed process and representation of state, input and output variables

The path leading to the model (2.5-45) could have been different. It is possible to divide the pipe into elements of finite volume first, and then formulate energy conservation equations for each of those elements separately, or for one typical element if all the subsystems are alike. This will be done in the following lines for an element between cross-sections n-1 and n and that derivation will be a suitable illustration of what discretization by the method of backward difference really represents, i.e. where an error is consciously made in the application of that method.

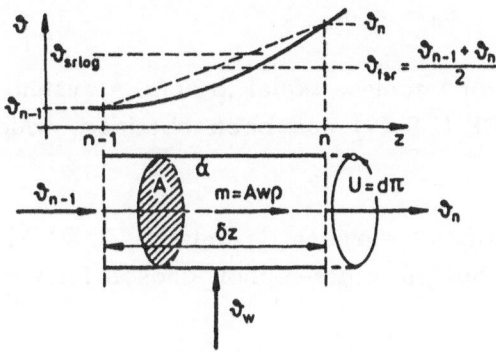

Fig. 2.5-5 Finite pipe element of the condenser heat exchanger

Figure 2.5-5 shows the (n-1)st element, for which an equation for the conservation of energy must be formulated. Above it is shown an exponential fluid temperature change along that element, given by Equation (3.1-29) in the steady state.

The energy conservation equation for the element is

$$mc\vartheta_{n-1} + \alpha U\delta z(\vartheta_w - \vartheta_{fsr}) - mc\vartheta_n = Mc\frac{d\vartheta_{fsr}}{dt} \quad . \tag{2.5-49}$$

ϑ_{fsr} denotes the mean fluid temperature between cross-sections n-1 and n, and the error resulting from discretization occurs in the selection of this temperature. It is possible to make many different selections and write

$$\vartheta_{fsr} = \vartheta_{n-1} \quad , \quad \text{or}$$

$$\vartheta_{fsr} = \frac{\vartheta_{n-1} + \vartheta_n}{2} \quad , \quad \text{or}$$

$\vartheta_{fsr} = \vartheta_n$, or

$\vartheta_{fsr} = \vartheta_{srlog}$.   ...   .

The thicker the division, the smaller the temperature difference between the cross-sections n-1 and n, which reduces the error introduced into discretization. At the limit when division into an infinite number of elements has been made, we have

$$\vartheta_{fsr} = \vartheta_{n-1} = \vartheta_n = \vartheta_{srlog}$$.

and the error becomes equal to zero. A system of an infinite number of first- order ODE (2.5-49) has been obtained, which is equivalent to PDE (2.5-43).

If the number of elements is finite (1, 3, 9, 17) and the temperature at the output of the n-th cross-section chosen for the mean fluid temperature, i.e.

$$\vartheta_{fsr} = \vartheta_n \tag{2.5-50}$$

the same system of ODE is obtained as the one derived by the method of backward difference and shown in Equation (2.5-45). Substitution of (2.5-50) into (2.5-49) for m = Awρ and M = Aρδz and shorter rearrangement yields the equation describing the dynamics of temperature of the (n-1)st element

$$\frac{d\vartheta_n}{dt} + w\frac{\vartheta_n - \vartheta_{n-1}}{\delta z} + k\vartheta_n = k\vartheta_w \tag{2.5-51}$$.

Therefore, the method of backward difference represents a substitution of mean fluid temperature in a finite element obtained through discretization, with the output fluid temperature of that element.

The mathematical discretization shown here can, as a rule, be applied to all the PDE that appear in this book. If it is, for example, applied to the system of three PDE (3.3-53) - (3.3-55), a system of 3· NE first-order ODE is obtained which can then be processed (simulated) on a digital computer.

Here we have completely left out an analysis of all the many problems linked with methods of discretization themselves, about which many specialized works, studies and books have been written. We must at least mention that this primarily concerns the accuracy of the model (in the sense of it approaching as closely as possible the real dynamics of the undiscretized system of PDE) its stability and size, which is connected with the costs of simulation and the difficulties of numerical processing.

This ends the description of dynamic models of processes with lumped parameters, and the last example is a transition to the modeling of processes in which some variables show spatial dependence as well. Most of them cannot be completely described by one or more ODE, so from now on partial DE will appear. This will be explained in more detail on the following pages.

DISTRIBUTED PROCESSES

In previous discussion we analyzed and described the dynamic properties of processes in which the dependence of state variables on the spatial coordinate was neglected, i.e. in which we considered that any change in state variable occurs simultaneously and equally in the whole part of the plant under observation. Strictly speaking, in by far the greatest number of cases that appear in engineering practice the velocities, temperature, flows, concentrations, forces, pressures and so on change differently both in time and in the separate parts of the plants. The dynamics of certain variables thus depends on the spatial coordinate and in these cases, when the time needed for disturbance propagation (changes in state variables) through the plant corresponds to the time constants characterizing other transient processes, that spatial distribution must not be neglected. There are many examples of such systems in which there is obvious spatial distribution and dependence of variables: pipe heat exchangers, gas pipes and pipelines, transport conveyers and so on, are examples in which state variables depend on one spatial coordinate (the spatial coordinate z), and the oscillation of membranes and shells, heating slabs, spreading of oil patches on the sea or forest fires are typical cases where there is process variable dependence on two spatial coordinates. (The last two cases have recently become especially interesting and their dynamics is the subject of intense study in many countries.) Finally, dynamic temperature changes in heating large casts (ingots) or the dynamics in steam generator evaporator pipes are today often analyzed by observing variable change in the whole volume of the cast or pipe, which means dependence on three spatial coordinates.

In the following text we have limited ourselves to processes that depend on only one spatial coordinate (z) and on time (t). For most technical devices and plants this approach completely satisfies the demands placed before the analysis of dynamic processes. It will be shown that even for those " simplest" cases of spatial dependence the mathematical apparatus and methodology of solution are of a higher degree of complexity and with less possibilities for general solution than was the case up to now.

Here too the description of dynamic phenomena and the solution of the partial differential equations (PDE) obtained from that description will be begun from the basic and simplest processes and mathematical structures, after which we will proceed to more complex phenomena and mathematical notation. In cases where complex and very extensive expressions and structures are not obtained, we will also show the analytical procedures for obtaining solutions of PDE.

At the very beginning of this third chapter it will be of advantage to remember some basic and general characteristics of PDE. They describe time changes of state variables (concentration, temperatures, flows, pressures, densities and so on) throughout the device or plant observed, i.e. in every arbitrary point in space, of which there are of course infinitely many. PDE therefore describe processes of infinite order. The fact that in practice a state variable change is important on the output (input) or in only several sections, and not in an infinite number of them, is not of great help in this case because state variable changes in the section under observation depend on the variable gradients with respect to the spatial coordinate, which makes it necessary to know the variables that are infinitesimally close to that section also. Analogously, for that infinitesimally close section the problem becomes (or remains) of infinitely high order. For everyday technical practice, therefore, the task of decreasing the model's order will be one of primary importance, and will as a rule be reduced to the possibility of accurately calculating the gradient of state variables with respect to the spatial coordinate with the help of a finite number of points.

The following must also be said in connection with PDE that will be the basic tools used to describe and analyze dynamic processes occurring in devices and plants along which parameters change. This part of the book is not intended to be, nor should it be considered, a shortened version or the repetition of books from mathematical physics or special chapters from thermodynamics, fluid mechanics and mechanics, in the sense that it gives answers and shows methods for classifying PDE, analyzes problems of existence and uniqueness of solutions, shows different methods for solving certain types of PDE and the like. Here, on many examples that are encountered in everyday technical practice, we will primarily try to indicate some basic dynamic properties and characteristics of processes distributed in space and show methods of deriving models for unsteady state changes. To do this, in analogy with lumped parameter processes, we will use Laplace transformation and transfer functions. Wishing every model obtained to be solvable and usable in practice, in many examples we will also show how a model of finite dimension is obtained, what is lost when the order of the model is reduced and how transcendental transfer functions are approximated by transfer functions in the form of proper rational function.

From the aspect of process linearity, the use of Laplace transformation and the formalization of symbols, we must repeat and bear in mind the remark that was already made at the beginning of Chapter 2. In all the following cases when the **PDE obtained are linear**, for the state variables observed they can also be considered as the **equations of the deviation** of those variables from some steady state. In the process of obtaining transfer functions, when the initial conditions are taken to equal zero this also means that the initial deviations equal zero, i.e. that the system is in steady state. Transfer functions obtained in this way are valid both for absolute values of variable change from the initial zero state (from initial conditions equal to zero), and also for changes of variable deviation from the initial steady state. In such linear cases this linearity will, therefore, not be specially stressed by introducing deviation symbol Δ beside the variable symbols (for example, we will not write Δm, $\Delta\rho$, $\Delta\vartheta$, and so on, but simply m, ρ, ϑ).

Finally, we must add that the mathematical model for these processes will also be derived from the **equations for the conservation of mass, energy and momentum**, but since the assumption of state variable equality throughout the whole finite volume observed no longer holds, these equations will be **formulated for the infinitesimally small part** of the volume dV for which the upper assumption is fulfilled.

3.1 MASS AND ENERGY TRANSPORTATION PROCESSES BASIC HYPERBOLIC FIRST-ORDER PARTIAL DIFFERENTIAL EQUATION

The transportation of bulk material on conveyer belts and the process of fluid flow through pipes, where heat energy is carried convectively from the pipe inlet to its outlet, are frequent, and the simplest examples of processes with spatially distributed parameters. Figures 3.1-1 and 3.1-2 show such processes of mass and energy transportation, and in the following lines we will show that the dynamic properties of those processes are described by the same PDE.

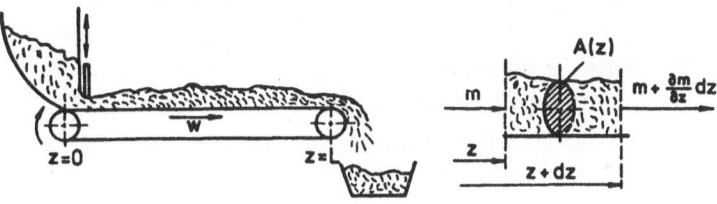

Fig. 3.1-1 Bulk material transportation

Fig. 3.1-2 Convective heat transfer by liquid flow

In deriving the model the assumptions for Figure 3.1-1 are :

- there is no side dissipation,

- the transport velocity w is constant.

The equation for the conservation of mass in the volume segment between z and dz is

$$m(z, t) - \left[m(z, t) + \frac{\partial m(z, t)}{\partial z} \, dz \right] = \frac{dM}{dt} \qquad (3.1\text{-}1)$$

Mass flow rate m and mass between z and dz are

$$m = A(z) \rho w \qquad (3.1\text{-}2)$$

$$dM = A(z) \rho \, dz \qquad (3.1\text{-}3)$$

The last equations give a **hyperbolic** first-order PDE

$$\frac{\partial m(z,t)}{\partial t} + w \frac{\partial m(z,t)}{\partial z} = 0 \qquad (3.1\text{-}4)$$

Before deriving the model that describes the propagation of heat disturbances along the pipe, the following assumptions are made:

- the pipe is insulated,
- the heat capacity of the pipe walls is neglected,
- the liquid flowing through the pipe is incompressible, i.e.
    $u = i - Pv = i = c\vartheta$,
- the flow velocity w is constant.

The equation for the conservation of energy for the fluid between z and dz is

$$u(z, t) - \left[u(z, t) + \frac{\partial u(z, t)}{\partial z} dz \right] = \frac{dU}{dt} \qquad (3.1\text{-}5)$$

Heat flow rate u and internal energy U are given by

$$u = mc\vartheta \qquad (3.1\text{-}6)$$

$$U = Mc\vartheta = A dz \rho c \vartheta \qquad (3.1\text{-}7)$$

Finally, the same hyperbolic first-order PDE is obtained

$$\frac{\partial \vartheta(z,t)}{\partial t} + w \frac{\partial \vartheta(z,t)}{\partial z} = 0 \qquad (3.1\text{-}8)$$

In both cases the steady state is determined by $\partial/\partial t = 0$, i.e. $\overline{m}$ = const. and $\overline{\vartheta}$ = const.

Equations (3.1-4) and (3.1-8) show that although the processes are in fact different, dynamically they are obviously the same. In both systems some variables (mass, heat energy) are transferred convectively along a spatial coordinate, so both processes are described by a **hyperbolic first-order PDE**. As a rule, only the amount of unloaded material, or the liquid temperature, in section $z = L$ is important, and if these variables are to be regulated (for the case w = const.) it is done by acting on m(0,t) or $\vartheta(0,t)$. It would thus be advantageous to find the relation between those two sections.

The discussion in Example 1 will be carried out for the case of bulk material transport, but if we want to get expressions for the case on Figure 3.1-2, it is enough to replace m by ϑ in the models obtained.

Before obtaining an expression for m(L,t), we must point out that the amount of material unloaded from $z = L$ does not depend only on m(0,t), i.e. on how much material is placed on the conveyer belt, but also on whether there was a lot, a little or no material already on it when it was put into operation. Mathematically expressed, it is necessary to know the initial conditions (IC) $m(z,t_0)$, i.e. the distribution of the material on the conveyer belt at the moment t_0. (Since we usually take $t_0 = 0$, zero will usually stand in the place of t_0 and this will describe the IC, for example m(z,0).)

Example 1 Conveyer system for bulk material transport. Time response

The process of bulk material transport on Figure 3.1-1 is described by the following PDE and the given boundary and initial conditions

$$\frac{\partial m(z,t)}{\partial t} + w\frac{\partial m(z,t)}{\partial z} = 0 \quad , \tag{3.1-9}$$

BC: $z = 0$, $m(0, t) = r(t)$, $t > t_0$

IC: $t = t_0$, $m(z,t_0) = p(z)$, $z \in (0, L)$

Velocity w = const., $r(t)$ and $p(z)$ are given functions of time and space.

a) Find the general analytical solution (or the solution in a closed form) and present it graphically for the following conditions

al) $r(t) = h = 1$ for $t \geq 0$, $p(z) = 0$,

a2) $r(t) = 0$, $p(z) = 1$, $0 \leq z \leq L$.

b) Find the transfer function $G(s) = M(L,s)/M(0,s) = M(L,s)/R(s)$.

The simplest way to obtain solutions for the hyperbolic first-order PDE is to use Laplace transformation that is carried out with respect to time t, so that the initial PDE is transformed into ODE with respect to z in the s-domain. Then the ODE obtained is solved in the s-domain by the usual procedures and its solution returned by inverse Laplace transform into the t-domain. There is

$$\mathcal{L}\left\{\frac{\partial m(z, t)}{\partial z}\right\} = \frac{dM(z, s)}{dz} = M'(z, s)$$

Application of the above transform to (3.1-9) yields

$$sM(z, s) - p(z) + wM'(z, s) = 0 \quad , \tag{3.1-10}$$

$$M'(z, s) = -\frac{s}{w}M(z, s) + \frac{p(z)}{w} \quad . \tag{3.1-11}$$

The last equation is a nonhomogeneous first-order ODE with respect to z in the s-domain, and it can be solved by the usual method of constants variation. The general solution will thus be the sum of the homogeneous and the particular solution

$$M(z, s) = M_h(z, s) + M_p(z, s) \quad , \tag{3.1-12}$$

$$M_h(z, s) = Ce^{-\frac{s}{w}z} \tag{3.1-12a}$$

$$M_p(z, s) = e^{-\frac{s}{w}z} \int_0^z \frac{1}{w} \, p(\xi) \, e^{\frac{s}{w}\xi} \, d\xi \tag{3.1-12b}$$

Substituting the expressions for M_h and M_p into (3.1-12) and establishing with the help of the BC $M(0,s) = R(s)$ that the constant $C = R(s)$, we get the final expression which must be returned into the time domain by inverse Laplace transform

$$M(z, s) = e^{-\frac{s}{w}z} \left[R(s) + \int_0^z \frac{1}{w} \, \dot{p}(\xi) \, e^{\frac{s}{w}\xi} \, d\xi \right] \ . \tag{3.1-13}$$

Denoting the expression in the brackets $F(s)$ and its original $f(t)$ yields

$$M(z, s) = e^{-\frac{s}{w}z} \, F(s) \ . \tag{3.1-14}$$

On the basis of the time translation theorem by which translation in the domain of the original function $f(t)$ corresponds to damping of function $F(s)$ in the s-domain (which is the case above), the inverse transform yields

$$m(z, t) = f(t - \frac{z}{w}) \tag{3.1-15}$$

However, $f(t)$ is still unknown and is obtained as follows

$$f(t) = \mathcal{L}^{-1}\{R(s) + F_1(s)\} = r(t) + \mathcal{L}^{-1}\{F_1(s)\} = \tag{3.1-16}$$

$$= r(t) + \mathcal{L}^{-1}\left\{ \int_0^z \frac{1}{w} \, p(\xi) \, e^{\frac{s}{w}\xi} d\xi \right\} = r(t) + \int_0^z \frac{1}{w} \, p(\xi) \underbrace{\mathcal{L}^{-1}\left\{ e^{\frac{\xi}{w}s} \right\}}_{\delta(t + \frac{\xi}{w})} \, d\xi \ .$$

The solution of the integral on the right-hand side follows

$$\int_0^z \frac{1}{w} \, p(\xi) \, \delta(t + \frac{\xi}{w}) \, d\xi \tag{3.1-17}$$

Substitution of $\xi = w\tau$ and $d\xi = wd\tau$ into (3.1-17) yields

$$\int_0^z p(w\tau) \delta(\tau + t) d\tau = p(-wt) \tag{3.1-18}$$

Inserting the last expression into (3.1-16) instead of the integral yields $f(t)$

$$f(t) = r(t) + p(-wt) \ . \tag{3.1-19}$$

According to Equation (3.1-15) the solution is translated, so the final expression for $m(z,t)$ is

$$m(z, t) = r(t - \frac{z}{w}) + p(z - wt) \quad . \tag{3.1-20}$$

Figures (3.1-3) and (3.1-4) are a graphical representation of the solution in three-dimensional space (m,z,t) for conditions a1) and a2)

Consequently, for $t > T_t$ there will be no more material on the conveyer, i.e. $m(z, t > T_t) = 0$.

b) By definition, transfer function G(s) represents the ratio of output signal transform to input signal transform, with initial conditions equal zero. Thus G(s) is not difficult to obtain. It is enough to substitute $p(\xi) = 0$ into Equation (3.1-13), after which the whole expression under the integral becomes equal to zero and remains

$$M(z, s) = e^{-\frac{z}{w}s} R(s) = M(0, s) e^{-\frac{z}{w}s} \tag{3.1-21}$$

If we insert z = L, we get

$$G(s) = \frac{M(L, s)}{M(0, s)} = e^{-\frac{L}{w}s} \tag{3.1-22}$$

Ratio L/w has the dimension of time and represents the time needed for a particle travelling at velocity w to move from z = 0 to z = L. It is usually given the subscript t (transported)

$$T_t = \frac{L}{w} \quad . \tag{3.1-23}$$

(This time is also often called "dead" time because it is a period of time during which nothing will appear in the section z = L if the IC equal zero, $m(L, t < T_t) = 0$, i.e. it is a period of time during which everything will be "dead" in the output section. See Figure 3.1-3.)

The next equation is the well-known transcendental function (and no longer in the form of a proper rational function) for transport processes

$$G(z, s) = \frac{M(z, s)}{M(0, s)} = e^{-T_t s} \quad . \qquad (T_t = \frac{z}{w}) \tag{3.1-24}$$

For the case of liquid flow where we seek connections between input and output temperatures, replacing M by θ in the last equation directly yields Equation (3.1-25).

a1) BC $r(t) = h = 1$ for $t \geq 0$ h ... Heaviside unit-step function
IC $p(z) = 0$, $0 \leq z \leq L$
$m(z, t) = h(t - \frac{z}{w})$

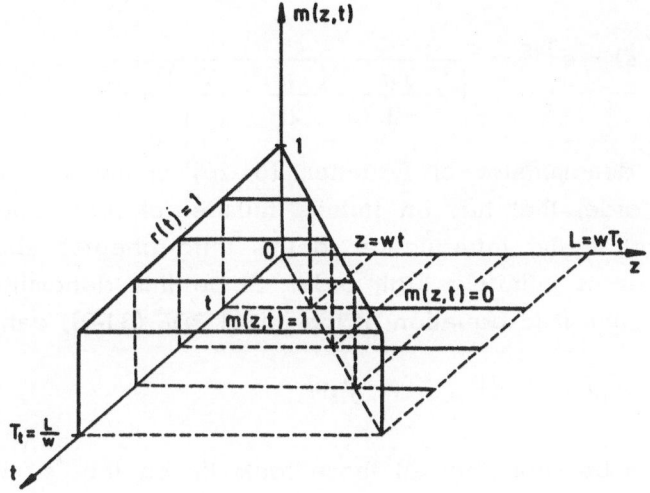

Fig. 3.1-3 Changes in quantity of material $m(z,t)$ for $r(t) = h$ and $p(z) = 0$

a2) BC $r(t) = 0$ for $t \geq 0$
IC $p(z) = 1$, $0 \leq z \leq L$
$m(z, t) = h(z - wt)$

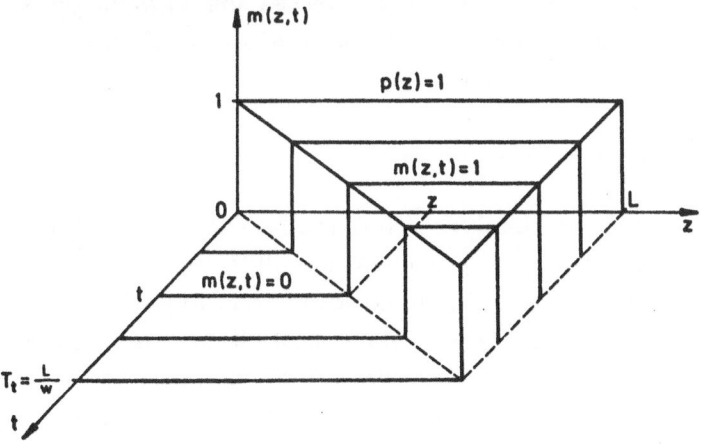

Fig. 3.1-4 Changes in quantity of material $m(z,t)$ for $r(t) = 0$ and $p(z) = 1$

$$G(z, s) = \frac{\Theta(z, s)}{\Theta(0, s)} = e^{-T_t s} \qquad\qquad (3.1\text{-}25)$$

If $e^{-T_t s}$ is expanded into an infinite series, the transfer function for transport processes obtains the following form

$$G(z, s) = e^{-T_t s} = \frac{1}{1 + \dfrac{T_t s}{1!} + \dfrac{(T_t s)^2}{2!} + \dfrac{(T_t s)^3}{3!} + \cdots} \qquad\qquad (3.1\text{-}26)$$

The denominator of Equation (3.1-26) contains a polynomial in s of infinite order that has an infinite number of roots and thus confirms the statement in the introduction to this third chapter about PDE describing processes of infinitely high order. Since this denominator (which is also the characteristic equation of the initial PDE (3.1-9)) can also be written

$$e^{T_t s} = e^{T_t(\sigma + \omega j)} = e^{T_t \sigma} e^{T_t \omega j} \quad ,$$

it can be seen that all those roots lie on the "straight line" $\sigma = -\infty$ in the complex s-plane and that they all have an arbitrary meaning for the imaginary part.

On the basis of everything that has been said, it follows clearly that **spatially distributed processes can not possibly be presented in the form of state-space equations**, which was always possible up to now for processes with lumped parameters. In this case the **dimension of the state vector x would be infinite**. It must also be said that in these processes the highest order of derivatives with respect to time does not equal the order of the system, which was always the case in lumped processes.

═══════════════

In Example 1 we analyzed the really simplest processes with spatial parameter dependence and with a selection of assumptions that "insured" a model in the form of a linear homogeneous hyperbolic first-order PDE. In the following example we will show how, if the assumptions are "marred" just a little (which is often closer to the real case), more complex mathematical forms are obtained to solve and analyze. (This change of assumptions could mean that material also slides during transport, or that it falls off the side of the conveyer belt, or that the pipe is not insulated, and so on.)

Example 2 Fluid flow through uninsulated pipe. Temperature time response

Derive a model for heat transfer by liquid flow through an uninsulated pipe, where the wall temperature is given by $\vartheta_w(z, t)$. Heat storage in the pipe wall is neglected and the mass flow is considered constant (and thus also flow velocity w). The consequence of the last assumption is that the coefficient of convective heat transfer α is constant as well.

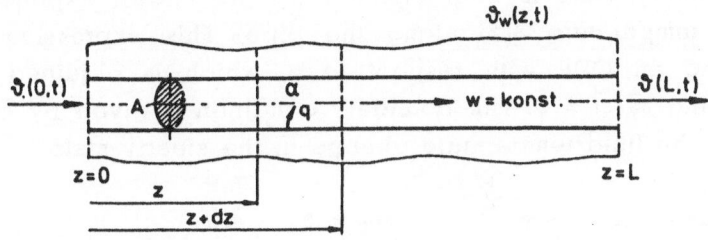

Fig. 3.1-5 Liquid flow through pipe and heat exchange with wall of temperature $\vartheta_w(z, t)$ ▆▆▆

This case is present in condenser heat exchangers and in their case the assumption of constant wall temperature, ϑ_w = const., is usually fulfilled.

Formulation of the equation for the conservation of energy for the fluid between sections z and z+dz yields

$$mc\vartheta - (mc\vartheta + mc\frac{\partial\vartheta}{\partial z}dz) + \alpha A_u(\vartheta_w - \vartheta) = \frac{d(Mc\vartheta)}{dt} \quad . \qquad (3.1-27)$$

The area through which heat is exchanged is $A_u = 2r\pi dz = Udz$, where U represents the circumference of the pipe. $M = A\rho dz$ is the mass of fluid in the observed dV. Arrangement of the last equation yields

$$\frac{\partial\vartheta(z, t)}{\partial t} + w\frac{\partial\vartheta(z, t)}{\partial z} + k\vartheta(z, t) = k\vartheta_w(z, t) \quad . \qquad (3.1-28)$$

$$k = \frac{\alpha U}{A\rho c}$$

The model is a nonhomogeneous linear hyperbolic first-order PDE and linear relations will be retained as long as the assumption about the constant flow velocity w is fulfilled. To obtain an analytic solution of Equation (3.1-28) it is necessary to know the boundary conditions function $\vartheta(0,t)$, the initial conditions function $\vartheta(z,0)$ and the disturbance function $\vartheta_w(z, t)$.

Coefficient k is an important dynamic coefficient and will be met in all pipe heat exchangers. The dimension of k is s^{-1} and its inverse value is called the **heat (thermal) time constant** of fluid T = 1/k.

For the case of the condenser heat exchanger where $\vartheta_w(z, t)$ = const., the steady state is characterized by the known exponential increase of fluid temperature $\vartheta(z)$ along the pipe. This expression is obtained by solving an ODE with respect to z, which is obtained from (3.1-28) by inserting $\partial\vartheta/\partial t = 0$. The boundary condition is given by $\bar{\vartheta}(z=0) = \bar{\vartheta}(0)$. This gives the fluid temperature change in the steady state

$$\bar{\vartheta}(z) = \bar{\vartheta}_w - \delta\bar{\vartheta}_0\, e^{-\frac{k}{w}z} \quad , \quad \delta\bar{\vartheta}_0 = \bar{\vartheta}_w - \bar{\vartheta}(0) \quad . \tag{3.1-29}$$

If $\bar{\vartheta}(0) = 0$

$$\bar{\vartheta}(z) = \bar{\vartheta}_w(1 - e^{-\frac{k}{w}z}) \quad , \tag{3.1-30}$$

and if both $\bar{\vartheta}(0) = 0$ and $\bar{\vartheta}_w(0) = 0$, the often used initial condition equal to zero is obtained, i.e. $\bar{\vartheta}(z) = 0$.

If we need to know the relation between fluid temperature change in an arbitrary section along the pipe and the input temperature (boundary condition) $\vartheta(0,t)$, or the temperature of the pipe wall $\vartheta_w(z, t)$, it is advantageous to determine the transfer functions that define that relation. Applying the Laplace transform to (3.1-28), for IC equal to zero, we get the following ODE with respect to z in the s-domain

$$w\frac{d\Theta(z, s)}{dz} + (s + k)\Theta(z, s) = k\Theta_w(z, s) \tag{3.1-31}$$

Solving that DE gives the demanded relation between temperatures

$$\Theta(z, s) = \underbrace{e^{-\frac{s+k}{w}z}}_{G_1(s)}\Theta(0, s) + \underbrace{\frac{1 - e^{-\frac{s+k}{w}z}}{s + k}k\Theta_w(z, s)}_{G_2(s)} \quad . \tag{3.1-32}$$

It is important, because of comparison with the case when the pipe is insulated as shown on Figures 3.1-3 and 3.1-4, to represent the response of the system (fluid temperature change at the pipe outlet) in the case when the inlet disturbance is an impulse fluid temperature change, $\vartheta(0, t) = \delta(t)$. (The wall temperature remains unchanged and equal to zero.) With $\theta(0, s)$ = 1 the above equation yields

$$\vartheta(z, t) = \mathscr{L}^{-1}\left\{ e^{-\frac{s+k}{w}z} \right\} = e^{-\frac{z}{w}k} \, \mathscr{L}^{1}\left\{ e^{-\frac{z}{w}s} \right\} = e^{-\frac{z}{w}k} \, \delta\left(t - \frac{z}{w}\right) . \qquad (3.1\text{-}33)$$

An impulse response for the convective heat transfer process (transfer by fluid flow) in which there is heat exchange with the wall is thus an impulse that propagates by velocity w along the pipe (with a time delay of z/w seconds after the occurrence of the disturbance) and whose amplitude decreases by the damping factor $e^{-zk/w}$.

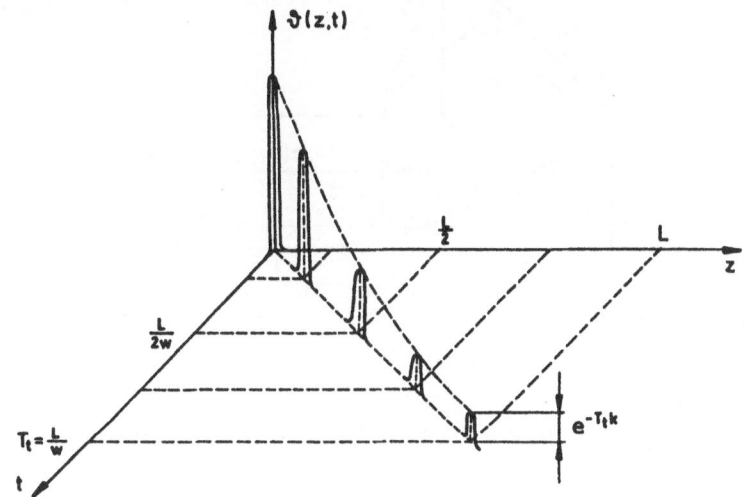

Fig. 3.1-6 Propagation of impulse temperature disturbance along the uninsulated pipe

Special attention must be paid to this decrease in amplitude shown on Figure 3.1-6 because it did not exist in the case of the insulated pipe. There the disturbance propagated along the pipe (transporter) unchanged. For better comparison with Figure 3.1-3 (which is also valid for the insulated pipe, as was mentioned in the first example) we should obtain an analytical expression for temperature change along the pipe if $\vartheta(0, t) =$

$r(t) = h(t) = 1$ for $t > 0$. Substituting $\Theta(0, s) = 1/s$ and $\Theta_w(z, s) = 0$ into (3.1-32) yields

$$\Theta(z, s) = e^{-\frac{z}{W}k}\, e^{-\frac{z}{W}s}\, \frac{1}{s} \quad .$$

whence

$$\vartheta(z, t) = e^{-\frac{z}{W}k}\, \mathcal{L}^{-1}\left\{ e^{-\frac{z}{W}s}\, \frac{1}{s}\right\} = e^{-\frac{z}{W}k}\, h(t - \frac{z}{W}) \qquad (3.1\text{-}34)$$

Figure 3.1-7 shows fluid temperature change.

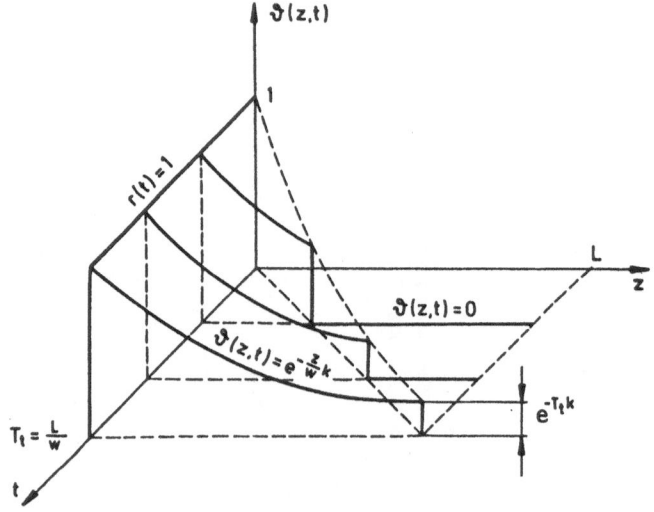

Fig. 3.1-7 Propagation of step temperature disturbance along the uninsulated pipe

The temperature profile along the pipe is shown at the moment T_t. Outlet temperature is equal to $e^{-Lk/w} = e^{-T_t k}$, which means that it is considerably smaller than inlet temperature.

—————————————

Pipe heat exchangers (parallel-flow, counter-flow and cross-flow) are frequent parts of plants and plant units and typical examples in which the dynamics of temperature change between the warmer and the colder current is described by a **system** of hyperbolic first-order PDE. In continuation, after accepting many assumptions, we will show how to derive a model for the parallel-flow and counter-flow heat exchanger. For the former we will also show how to obtain the matrix transfer function relating input and output temperatures.

═══════════════

Example 3 Dynamics of parallel-flow and counter-flow heat exchangers

Figure 3.1-8 shows a sketch of counter-flow and parallel-flow (the direction of flow is given by a broken line) heat exchangers for which mathematical models for unsteady fluid temperature changes must be derived. The assumptions are:

- both fluids are incompressible liquids,

- pressures, mass flow rates, flow velocities, cross-sectional areas and heat transfer coefficients are constant,

- the heat capacitance of the pipe walls, heat conduction in the direction of the z axis and losses due to friction are neglected,

- the coefficient of heat conduction through the pipe wall is infinite.

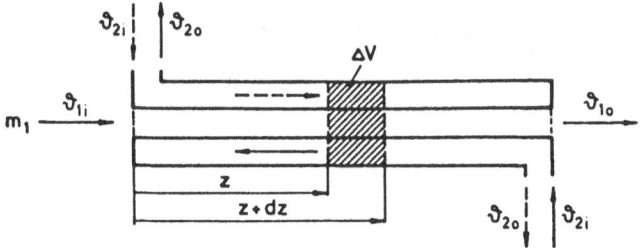

Fig. 3.1-8 Schematic presentation of counter-flow and parallel-flow heat exchanger

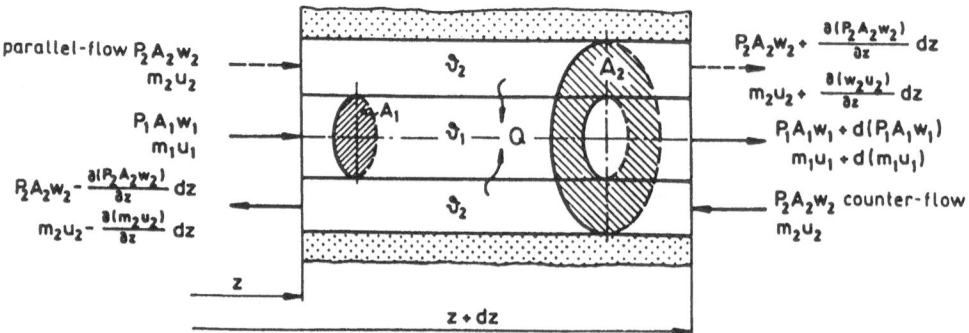

Fig. 3.1-9 Sketch of control volume dV with heat inflow and heat outflow

Equations for the conservation of energy (power equations) are formulated for infinitesimal volumes of both the fluids between sections z and z+dz.

ECE for fluid 1:

$$m_1u_1 + P_1A_1w_1 + Udzq - (m_1u_1 + \frac{\partial(m_1u_1)}{\partial z}dz) - (P_1A_1w_1 + \frac{\partial(P_1A_1w_1)}{\partial z}dz) =$$

$$= \frac{\partial}{\partial t}(A_1dz\rho_1u_1) \qquad\qquad\qquad (3.1\text{-}35)$$

yields

$$\frac{\partial}{\partial z}(m_1u_1 + P_1A_1w_1) + \frac{\partial}{\partial t}(A_1\rho_1u_1) = Uq \qquad\qquad (3.1\text{-}36)$$

Considering that

$$m_1u_1 + P_1A_1w_1 = m_1u_1 + P_1v_1m_1 = m_1(u_1 + P_1v_1) = m_1i_1 \quad .$$

$$\rho_1u_1 = \rho_1i_1 - P. \ A_1 \frac{\partial\rho_1}{\partial t} = \frac{\partial m_1}{\partial z} , \ P = const. \quad .$$

differentiation of Equation (3.1-36) yields

$$\frac{\partial i_1}{\partial t} + w_1\frac{\partial i_1}{\partial z} = \frac{U}{A_1\rho_1}q = \frac{U}{A_1\rho_1}\alpha(\vartheta_2 - \vartheta_1) \quad . \qquad\qquad (3.1\text{-}37)$$

For liquids there is $i = c\vartheta$, so the last equation yields the final expression for fluid 1 temperature change

$$\frac{\partial \vartheta_1(z, t)}{\partial t} + w_1 \frac{\partial \vartheta_1(z, t)}{\partial z} = \frac{U\alpha}{A_1 \rho_1 c_1}(\vartheta_2 - \vartheta_1) \; . \tag{3.1-38}$$

The ECE for the warmer fluid 2 (the upper signs are valid for the parallel-flow heat exchanger) is

$$m_2 u_2 + P_2 A_2 w_2 - Udzq - (m_2 u_2 \pm \frac{\partial (m_2 u_2)}{\partial z} dz) - (P_2 A_2 w_2 \pm \frac{\partial (P_2 A_2 w_2)}{\partial z} dz) =$$

$$= \frac{\partial}{\partial t}(A_2 dz \rho_2 u_2) \tag{3.1-39}$$

In this equation attention must be paid to the negative sign in front of the heat transferred to fluid 1, and in front of the partial derivatives with respect to z in the counter-flow exchanger. These signs are what distinguishe Equation (3.1-39) (for the heating fluid) from Equation (3.1-35) (for the heated fluid) and they always reflect the fact that fluid 2 gives heat to fluid 1, and that the energy level of fluid 2 in the counter-flow exchanger decreases in the negative direction of the z axis.

If the same procedure as the one carried out with the equation for fluid 1 is performed, we get

$$\frac{\partial \vartheta_2(z, t)}{\partial t} \pm w_2 \frac{\partial \vartheta_2(z, t)}{\partial z} = \frac{U\alpha}{A_2 \rho_2 c_2}(\vartheta_1 - \vartheta_2) \tag{3.1-40}$$

Equations (3.1-38) and (3.1-40) are a **system** of hyperbolic first-order PDE, and to solve them it is necessary to define **two boundary** and **two initial conditions**. In the counter-flow exchanger these conditions are of the form

BC: $\vartheta_1(0, t) = r_1(t)$, $\vartheta_2(L, t) = r_2(t)$.
IC: $\vartheta_1(z, t_0) = p_1(z)$, $\vartheta_2(z, t_0) = p_2(z)$.

The upper assumptions make the system of equations homogeneous, the process occurs without disturbance and only because the boundary conditions change. If the matrix differential operator $\mathbf{A}_z$ is defined and the coefficients $k_i = U\alpha/A_i \rho_i c_i$, $i = 1, 2$ are reintroduced, the system of equations can be written as follows

$$\vartheta_t'(z, t) = \begin{bmatrix} -w_1 \frac{\partial}{\partial z} - k_1 & k_1 \\[2mm] k_2 & \mp w_2 \frac{\partial}{\partial z} - k_2 \end{bmatrix} \vartheta(z, t) = \mathbf{A}_z \vartheta(z, t) \tag{3.1-41}$$

Obtaining closed form solutions for this system of equations is now a rather more considerable task than when there was only one equation. The process has increased in complexity. Here we will, nevertheless, show how to obtain the matrix transfer function which relates temperatures along the **parallel-flow** heat exchanger as functions of boundary conditions: temperatures $\vartheta_1(0, t)$ and $\vartheta_2(0, t)$.

With IC equal to zero, after Laplace transformation with respect to time t the system of equations (3.1-41) becomes the following system of ODE with respect to z in the s-domain

$$\frac{d\theta_1(z, s)}{dz} = -\frac{s + k_1}{w_1} \theta_1(z, s) + \frac{k_1}{w_1} \theta_2(z, s) \quad , \tag{3.1-42}$$

$$\frac{d\theta_2(z, s)}{dz} = \frac{k_2}{w_2} \theta_1(z, s) - \frac{s + k_2}{w_2} \theta_2(z, s) \tag{3.1-43}$$

To make it easier to perform the following derivation we will assume that the **fluids are the same** ($\rho_1 = \rho_2$, $c_1 = c_2$) and that $A_1 = A_2$ and $w_1 = w_2 = w$. This results in $k_1 = k_2 = k$. After introducing the symbols $A = -(s+k)/w$ and $B = k/w$, the above system of equations shown in matrix form becomes

$$\begin{bmatrix} \theta_1(z, s) \\ \theta_2(z, s) \end{bmatrix}_z' = \begin{bmatrix} A & B \\ B & A \end{bmatrix} \begin{bmatrix} \theta_1(z, s) \\ \theta_2(z, s) \end{bmatrix} \tag{3.1-44}$$

$$\mathbf{A}$$

The solution of system (3.1-44) is

$$\Theta(z, s) = e^{\mathbf{A}z} \Theta(0, s) = \Phi(z) \Theta(0, s) \tag{3.1-45}$$

The transient or fundamental matrix $\Phi(z)$ is calculated according to the formula

$$\Phi(z) = e^{\lambda_1 z} \frac{\mathbf{A} - \lambda_2 \mathbf{I}}{\lambda_1 - \lambda_2} + e^{\lambda_2 z} \frac{\mathbf{A} - \lambda_1 \mathbf{I}}{\lambda_2 - \lambda_1} \quad . \tag{3.1-46}$$

λ_1 and λ_2 are the eigenvalues of matrix $\mathbf{A}$ obtained from the equatio ns $(\mathbf{A} - \lambda \mathbf{I}) = 0$.

$$\lambda_1 = A + B = -\frac{s}{w} \quad , \tag{3.1-47}$$

$$\lambda_2 = A - B = -\frac{s + 2k}{w} \quad . \tag{3.1-48}$$

Equation (3.1-46), after determining matrices $\mathbf{F}_1$ and $\mathbf{F}_2$, yields

$$\boldsymbol{\Phi}(z) = e^{Az}\begin{bmatrix} chBz & shBz \\ shBz & chBz \end{bmatrix} = \frac{e^{Az}}{2}\begin{bmatrix} e^{Bz}+e^{-Bz} & e^{Bz}-e^{-Bz} \\ e^{Bz}-e^{-Bz} & e^{Bz}+e^{-Bz} \end{bmatrix} \qquad (3.1\text{-}49)$$

Temperature interrelations can be shown clearly in a block diagram, and this has been done in Figure 3.1-10 according to Equations (3.1-45) and (3.1-49).

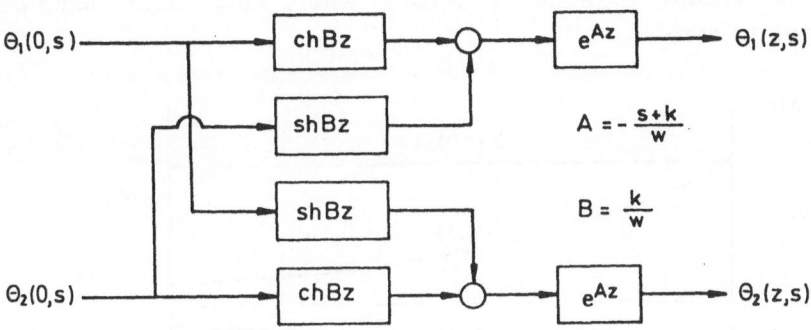

Fig. 3.1-10 Block diagram of interrelations between input and output temperatures in the parallel-flow heat exchanger
$(k_1 = k_2 = k, \ w_1 = w_2 = w)$

Equations (3.1-45) and (3.1-49) make it possible to determine temperatures along the whole heat exchanger for given temperature changes $\vartheta_1(0, t)$ and $\vartheta_2(0, t)$. If ϑ_1 undergoes unit step increase, i.e. $\vartheta_1(0, t) = h(t)$, the temperature change of both fluids is given by the solution of the following equation

$$\begin{bmatrix} \Theta_1(z, s) \\ \Theta_2(z, s) \end{bmatrix} = \frac{e^{-\frac{s+k}{w}z}}{2}\begin{bmatrix} e^{\frac{k}{w}z}+e^{-\frac{k}{w}z} & e^{\frac{k}{w}z}-e^{-\frac{k}{w}z} \\ e^{\frac{k}{w}z}-e^{-\frac{k}{w}z} & e^{\frac{k}{w}z}+e^{-\frac{k}{w}z} \end{bmatrix}\begin{bmatrix} \frac{1}{s} \\ 0 \end{bmatrix} \qquad (3.1\text{-}50)$$

Rearrangement and inverse Laplace transformation yields

$$\vartheta_1(z, t) = \frac{1+e^{-2\frac{z}{w}k}}{2}\, h(t - \frac{z}{w}) \quad . \qquad (3.1\text{-}51)$$

$$\vartheta_2(z, t) = \frac{1 - e^{-2\frac{z}{w}k}}{2} \ h(t - \frac{z}{w}) \quad , \tag{3.1-52}$$

For the case of $k = 1$, pipe length $L = 1$ m and flow velocity $w = 1$ m/s, Figure 3.1-11 shows temperature changes of both the fluids along the whole exchanger after $T_t = L/w = 1$ second. It must be observed that when $z \rightarrow \infty$, ϑ_1 and ϑ_2 converge to the value 0.5, and this differs from the results in Figure 3.1-7 (ϑ_w = const.) where the fluid temperature ϑ converges to zero.

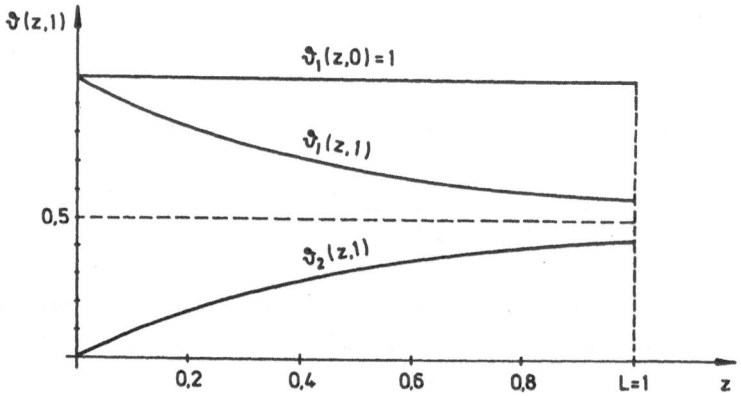

Fig. 3.1-11 Temperature profile along heat exchanger pipe after T_t seconds in the case $\vartheta_1(0, t) = h(t) = 1$

This last example shows how technical devices with relatively simple geometry and for which a whole series of assumptions has been made, have complicated mathematical models which make it difficult to obtain analytically the demanded responses in the case of disturbed boundary conditions. Therefore it, and also the many examples that follow, shows why everyday technical practice aspires to methods of direct numerical simulation of PDE in the examination of dynamic properties of specific processes, or to approximation of transcendental transfer functions which always appear in processes that are distributed in space.

In the previous discussion of heat exchangers the possibility of heat accumulation in the walls of the pipe was neglected. In a great number of cases for thin-walled pipes this is completely justified. However, when the wall of the pipe is thicker, or when the heat capacitance of the wall is of the same order of magnitude as the heat capacitance of the fluid (which the model builder must evaluate at the beginning), the process of heat storage in the pipe wall must be taken into account as well. The following example shows changes in the model of unstationary temperature change for that case.

Example 4 Fluid flow with heat storage in the pipe wall

Fluid flows through a thicker pipe, externally insulated. Derive the transfer function relating output fluid temperature change with input fluid temperature change, taking into account possibilities of heat storage in the wall. Assume that the coefficient of heat conduction in the radial direction is infinite, which means that the temperature of the wall is the same in its whole thickness. Neglect heat conduction in the direction of the z axis. The other assumptions are the same as in the preceding example.

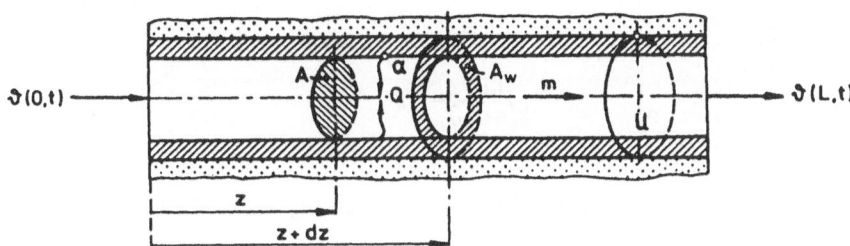

Fig. 3.1-12 Convective heat transfer through an insulated pipe with heat storage in the pipe wall

Because of the appearance of a new possible heat storage tank equations for the conservation of energy both for the fluid and for the pipe wall must now be derived.

ECE for the fluid:

$$mc\vartheta - \left(mc\vartheta + \frac{\partial(mc\vartheta)}{\partial z}dz\right) - \alpha U dz(\vartheta - \vartheta_w) = \frac{\partial(Adz\rho c\vartheta)}{\partial t} \qquad (3.1\text{-}53)$$

The above equation yields

$$\frac{\partial\vartheta}{\partial t} + w\frac{\partial\vartheta}{\partial z} = \frac{\alpha U}{A\rho c}(\vartheta_w - \vartheta) \quad , \qquad (3.1\text{-}54)$$

ECE for the wall:

$$A_w\rho_w c_w dz\frac{\partial\vartheta_w}{\partial t} = \alpha U dz(\vartheta - \vartheta_w) \qquad (3.1\text{-}55)$$

$$\frac{\partial\vartheta_w}{\partial t} = \frac{\alpha U}{A_w\rho_w c_w}(\vartheta - \vartheta_w) \qquad (3.1\text{-}56)$$

The system comprising Equations (3.1-54) and (3.1-56) is a model for unsteady fluid and wall temperature changes and to solve it we must define the following boundary and initial conditions:

BC: $\vartheta(0, t) = r(t)$
IC: $\vartheta(z, 0) = p_1(z)$, $\vartheta_w(z, 0) = p_2(z)$

The transfer function relating fluid temperature along the pipe with the given boundary temperature at $z = 0$ is obtained after Laplace transformation of Equations (3.1-54) and (3.1-56) with IC equal to zero, and after solving the ODE obtained with respect to z in the s-domain for fluid temperature ϑ.

$$G(z, s) = \frac{\Theta(z, s)}{\Theta(0, s)} = e^{-\frac{s^2+(k_1+k_2)s}{w(s+k_2)}z} \qquad (3.1\text{-}57)$$

The coefficients k have the same meaning as before

$$k_1 = \frac{\alpha U}{A\rho c} \quad , \qquad (3.1\text{-}58)$$

$$k_2 = \frac{\alpha U}{A_w\rho_w c_w} \quad . \qquad (3.1\text{-}59)$$

Their inverse values are fluid thermal time constant T_F and wall thermal time constant T_w.

$$T_F = \frac{1}{k_1} = \frac{A\rho c}{\alpha U} \quad , \qquad (3.1\text{-}60)$$

$$T_w = \frac{1}{k_2} = \frac{A_w \rho_w c_w}{\alpha U} \quad . \tag{3.1-61}$$

The transfer function $G(z,s)$ given in Equation (3.1-57) is a more encompassing and general form of transfer function for the pipe heat exchanger. From it we can obtain as boundary cases the transfer functions derived earlier if we make assumptions that were left out in this case. Thus, for example, if we allow the coefficient α to converge to zero, i.e. if there is no heat transfer onto the wall, the transfer function (3.1-57) becomes the already known transfer function that was derived for the "pure" transport process in the case of the insulated pipe in Example 1, given by Equation (3.1-25).

―――――――――

This ends the presentation of the dynamics of convective mass or energy transfer, better known as transport processes, whose unsteady state changes we described in the form of a hyperbolic firstorder PDE or by a system of hyperbolic first-order PDE. These are also the simplest processes distributed in space and their analysis will be of use for understanding more complex phenomena.

3.2 PROCESSES WITH EQUALIZATION
BASIC PARABOLIC PARTIAL DIFFERENTIAL EQUATION

This name describes processes in which differences in potential result in an equalization of conditions throughout the system under observation. The most typical examples met in everyday practice are heat conduction, diffusion and the motion of viscous (sticky) liquids. Common to all these physically different processes is again their mathematical notation: their dynamics is described by **parabolic PDE**. In this section we will completely analyze the process of heat conduction, but the derivation of equations, methods for solving them and the solutions themselves are identical for this whole class of problems, and to obtain a satisfactory physical interpretation it is enough to give coefficients their real meaning.

If certain assumptions are made, it can be shown that all the processes mentioned are described by the following PDE,

$$\frac{\partial u}{\partial t} = a\frac{\partial^2 u}{\partial z^2} \quad . \tag{3.2-1}$$

Depending on the nature of the process, u and a have the following physical meaning:

Heat conduction: u – temperature, a = $\lambda/c\rho$ – coefficient of thermal conductivity

diffusion: u – concentration, a = D – coefficient of diffusion

motion of viscous liquid: u – particle velocity, a = ν – coefficient of kinematic viscosity

Of course, as assumptions change so does Equation (3.2-1). It gets new terms and expands. Nevertheless, the form it is shown in is basic for all processes with equalization.

═══════════════════════════

Example 1 Heat conduction through homogeneous insulated body

Figure 3.2-1 shows a prismatic body insulated on all its lateral sides. Derive the equation describing temperature change along the spatial coordinate and in time, and express mathematically the following boundary conditions:

a) given temperatures are maintained at the ends of the body (frontal cross-sections),

b) a given heat flow rate is externally supplied to the ends of the body,

c) there is convective heat transfer from the ends of the body to the environment.

Also derive transfer functions characterizing heat conduction for the boundary conditions for case a), and for a combination of boundary conditions for cases a) and b).

Assumptions:

- temperature is uniform on a cross-section of the body,
- the body is homogeneous,
- heat conduction is observed along the z axis only,
- the coefficient of thermal conductivity λ is not a function of temperature.

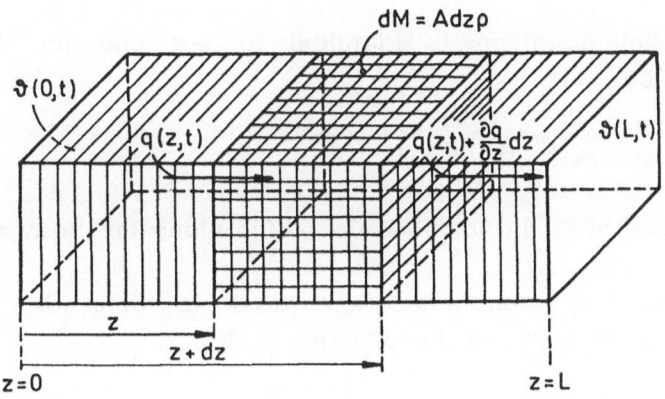

Fig. 3.2-1 Heat conduction through homogeneous insulated body

The balance of heat energy is formulated for a infinitesimal volume dV of the body

$$q(z, t)A - \left[q(z, t) + \frac{\partial q(z, t)}{\partial z} dz\right]A = \frac{dU}{dt} = A\rho c \frac{\partial \vartheta(z, t)}{\partial t} dz \qquad (3.2\text{-}2)$$

Rearrangement of (3.2-2) yields

$$-\frac{\partial q(z, t)}{\partial z} = \rho c \frac{\partial \vartheta(z, t)}{\partial t} \qquad (3.2\text{-}3)$$

The density of heat flow is known to be proportional to the spatial gradient of temperature and is expressed as follows (Fourier's law)

$$q(z, t) = -\lambda \frac{\partial \vartheta(z, t)}{\partial z} \quad . \qquad (3.2\text{-}4)$$

Substitution of (3.2-4) into (3.2-3) and differentiation yield

$$\frac{\partial \vartheta(z, t)}{\partial t} = a \frac{\partial^2 \vartheta(z, t)}{\partial z^2} \qquad (3.2\text{-}5)$$

Consequently, the dynamics of temperature change along the body is described by the linear parabolic PDE (3.2-5).

If we want to know the temperature in a specific cross-section at a given moment we must know the temperature distribution at $t_0 = 0$, which is considered the beginning of calculation (i.e. we must know the initial conditions), and also the conditions at the edges (boundaries) of the body (boundary conditions), i.e. how the body "contacts" its environment.

The initial condition is identical to the one for the hyperbolic first-order PDE

IC: $\vartheta(z, t_0) = p(z)$, $0 < z < L$

and it describes the temperature profile along the body at t_0.

The boundary conditions, according to demands made at the beginning of the task, can be of three kinds:

a) $\vartheta(0, t) = f_1(t)$, $\vartheta(L, t) = f_2(t)$.   (3.2-6)

b) $-\lambda A \dfrac{\partial \vartheta(0, t)}{\partial z} = q_1(t)$, $\lambda A \dfrac{\partial \vartheta(L, t)}{\partial z} = q_2(t)$.   (3.2-7)

c) $\dfrac{\partial \vartheta(0, t)}{\partial z} = \dfrac{\alpha}{\lambda}\left[\vartheta(0, t) - f_1(t)\right]$, $\dfrac{\partial \vartheta(L, t)}{\partial z} = -\dfrac{\alpha}{\lambda}\left[\vartheta(L, t) - f_2(t)\right]$

                                                     (3.2-8)

The functions $p(z)$, $f_1(t)$, $f_2(t)$, $q_1(t)$ and $q_2(t)$ are given. $f_1(t)$ and $f_2(t)$ are temperatures of the environment at the boundaries $z = 0$ and $z = L$, and $q_1(t)$ and $q_2(t)$ are heat flow rates (i.e. the amount of heat in a unit of time) forced on the ends of the body. It is obvious, but should nevertheless be mentioned, that besides the three above pairs of boundary conditions, their combinations are also possible, for example, temperature $f_1(t)$ can be given in $z = 0$, and convective heat exchange in $z = L$, i.e. boundary condition c). a), b) and c) do not show all the ways in which heat can be transferred to the end cross-sections. For example, there can also be radiation or, if there is firm contact with some other surface at $z = 0$ and $z = L$ combined with low thermal resistance, heat will be transferred by conduction.

The boundary conditions shown by c) are the most general and conditions a) and b) can arise as boundary cases of Equation (3.2-8).

If the convective heat transfer coefficient α is very large, theoretically even infinite, in other words if it is much greater than λ, condition c) becomes boundary condition a) because $\alpha \, \check{c} \, \infty$ leads to $\vartheta(0, t) = f_1(t)$ and $\vartheta(L, t) = f_2(t)$. But if α is insignificant, in other words if heat transfer on the edges is completely impossible, condition c) becomes boundary condition b) for $q_1(t) = q_2(t) = 0$, which really satisfies the case of insulated frontal cross-sections.

The transfer function for the condition $p(z) = 0$ will again be obtained after Laplace transformation of Equation (3.2-5) with respect to time t and solving the ODE obtained with respect to z

$$\frac{d^2\theta(z, s)}{dz^2} - \frac{s}{a}\,\theta(z, s) = 0 \qquad\qquad (3.2\text{-}9)$$

Solving Equation (3.2-9) yields

$$\theta(z, s) = C_1 e^{-\sqrt{\frac{s}{a}}z} + C_2 e^{\sqrt{\frac{s}{a}}z} \quad . \qquad\qquad (3.2\text{-}10)$$

The law according to which temperature changes along the z axis, i.e. the transfer function that shows that law, obviously depends only on conditions at the boundary of the body. Thus C_1 and C_2 depend only on the manner in which the process occurs at the boundaries, so they must be specially determined for the conditions given at the beginning of this example.

TEMPERATURE IS GIVEN AT BOTH ENDS

Boundary conditions: $\vartheta(0, t) = f_1(t)$, $z = 0$ (left end) $\vartheta(L, t) = f_2(t)$, $z = L$ (right end)

Substituting these boundary conditions into (3.2-10) yields the expressions for C_1 and C_2

$$C_1 = \frac{e^{\sqrt{\frac{s}{a}}L}}{2\,\mathrm{sh}(\sqrt{\frac{s}{a}}\,L)}\,F_1(s) - \frac{1}{2\,\mathrm{sh}(\sqrt{\frac{s}{a}}\,L)}\,F_2(s) \quad . \qquad\qquad (3.2\text{-}11)$$

$$C_2 = \frac{-e^{-\sqrt{\frac{s}{a}}L}}{2\,sh(\sqrt{\frac{s}{a}}\,L)}\,F_1(s) + \frac{1}{2\,sh(\sqrt{\frac{s}{a}}\,L)}\,F_2(s) \; . \qquad\qquad (3.2\text{-}12)$$

Inserting C_1 and C_2 into (3.2-10) and substituting boundary temperatures $\theta(0,\ s)$ and $\theta(L,\ s)$ for $F_1(s)$ and $F_2(s)$ yields

$$\theta(z,\ s) = \frac{sh\left[\sqrt{\frac{s}{a}}(L\text{-}z)\right]}{sh(\sqrt{\frac{s}{a}}L)}\,\theta(0,\ s) + \frac{sh(\sqrt{\frac{s}{a}}z)}{sh(\sqrt{\frac{s}{a}}L)}\,\theta(L,\ s) \qquad\qquad (3.2\text{-}13)$$

or

$$\theta(z,\ s) = G_1(z,\ s)\,\theta(0,\ s) + G_2(z,\ s)\,\theta(L,\ s)$$

TEMPERATURE GIVEN AT LEFT END, RIGHT END INSULATED

Boundary conditions:

$$\vartheta(0,\ t) = f_1(t), \quad z = 0$$

$$\frac{\partial\vartheta(L,\ t)}{\partial z} = 0, \quad z = L$$

The coefficients C_1 and C_2 can be obtained from Equation (3.2-10) and its differentiation with respect to z (to satisfy the second boundary condition). It is not difficult to prove that

$$C_1 = \frac{e^{\sqrt{\frac{s}{a}}L}}{2\,ch(\sqrt{\frac{s}{a}}\,L)}\,\theta(0,\ s) \qquad\qquad (3.2\text{-}14)$$

$$C_2 = \frac{e^{-\sqrt{\frac{s}{a}}L}}{2\,ch(\sqrt{\frac{s}{a}}\,L)}\,\theta(0,\ s) \qquad\qquad (3.2\text{-}15)$$

Equation (3.2-10), referring to (3.2-14) and (3.2-15), yields

$$G_3(z,\ s) = \frac{\theta(z,\ s)}{\theta(0,\ s)} = \frac{ch\left[\sqrt{\frac{s}{a}}(L\text{-}z)\right]}{ch(\sqrt{\frac{s}{a}}L)} \qquad\qquad (3.2\text{-}16)$$

It must be observed that the equations for transfer function (3.2-13) and (3.2-16) are functions of both variable z and variable s. There is an unlimited number of such transfer functions, and it is possible, using the corresponding z, to determine for every cross-section the relation between its temperature and the given boundary conditions.

The transfer function (3.2-16) will be of use to analyze the following example, which is typical for a whole class of process devices - direct heat exchangers with a thick wall.

Example 2 Direct heat exchanger with a thick wall

Figure 3.2-2 shows an **externally insulated** tank with a thick wall in which a liquid is heated and intensely mixed, insuring its uniform temperature throughout the whole volume. The process demands that the output temperature must be controlled. For constant input and output m, the only disturbance variable is the input fluid temperature ϑ_{fi}. The heat supplied by the electrical heater Q_{el} is in this case a manipulating (controling) variable. Determine the transfer functions relating liquid temperature ϑ in the tank with input temperature ϑ_{fi} and heat Q_{el}.

Assumptions: perfect and vigorous mixing so that the coefficient of heat transfer onto the wall can be considered very large ($\alpha \, \check{c} \, \infty$) or, which is the same, the wall temperature on the inner surface can be considered to equal the fluid temperature in the tank $\vartheta_w(0, t) = \vartheta(t)$.

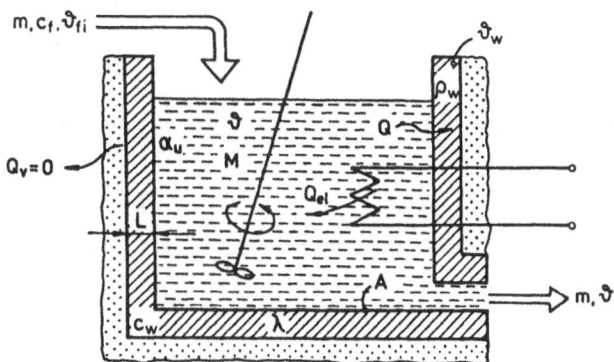

Fig. 3.2-2 Heater with a thick wall (direct heat exchanger)

The equation for the conservation of (heat) energy yields

$$mc_f \vartheta_{fi} + Q_{el} - mc_f \vartheta - Q = Mc_f \frac{d\vartheta}{dt} \tag{3.2-17}$$

Q is the heat flow rate on the tank wall and it can be expressed as the time change in heat energy stored in the wall

$$Q = \frac{dE}{dt} \tag{3.2-18}$$

Knowing the law governing the temperature change of the wall $\vartheta_w(z, t)$ along z, E can be expressed as

$$E = \int_0^L A\rho_w c_w \vartheta_w(z, t) \, dz \tag{3.2-19}$$

Substituting Equation (3.2-19) into (3.2-18) yields

$$Q(t) = \frac{d}{dt} \int_0^L A\rho_w c_w \vartheta_w(z, t) \, dz \tag{3.2-20}$$

If we consider that $E(0) = 0$ at $t = 0$, Laplace transformation of (3.2-20) yields

$$Q(s) = s \int_0^L A\rho_w c_w \Theta_w(z, s) \, dz \tag{3.2-21}$$

Inserting $\Theta_w(z, s)$ from Equation (3.2-16) into (3.2-21) gives the final expression

$$Q(s) = \frac{A\rho_w c_w s}{\sqrt{\frac{s}{a}}} \, \text{th}(\sqrt{\frac{s}{a}} L) \Theta(s) \tag{3.2-22}$$

Transformation of Equation (3.2-17) into the s-domain, substitution of (3.2-22) for $Q(s)$ and shorter rearrangement yield

$$\Theta(s) = G(s)\Theta_{fi}(s) + \frac{1}{mc_f} G(s)Q_{el}(s) \tag{3.2-23}$$

$$G(s) = \frac{1}{\frac{M}{m}s + \frac{A\rho_w c_w s}{\sqrt{\frac{s}{a}}\, mc_f}\, th(\sqrt{\frac{s}{a}}L) + 1} \qquad (3.2\text{-}24)$$

The transfer function G(s) completely determines changes in the fluid temperature ϑ in the tank for arbitrary changes of ϑ_{fi} and Q_{el}.

This example has not been chosen only to show that transcendental forms appear in G(s). (That this is so in spatially distributed systems has probably already been accepted and is clear.) Here we want to show that the transfer function (3.2-24) is a general form that comprises (as boundary cases) transfer functions that were obtained when this heater was observed as a system with lumped parameters (Equations (2.3-32) and (2.3-40)).

a) The tank walls are made of an insulating material for which $\lambda = 0$. In that case

$$G(s) = \frac{1}{\frac{M}{m}s + \underbrace{\frac{A\rho_w c_w s}{\sqrt{\frac{\rho_w c_w}{\lambda}s}\, mc_f}\, th(\sqrt{\frac{\rho_w c_w}{\lambda}s}\, L)}_{0} + 1} = \frac{1}{\frac{M}{m}s + 1}$$

$$(3.2\text{-}25)$$

Equation (3.2-23) has thus become (2.3-32).

b) The tank walls are made of a material with very great thermal conductivity ($\lambda \to \infty$ and $\vartheta_w(z, t) = \vartheta(t)$, $0 < z < L$).

In this case, when $\lambda \to \infty$, it can be shown that the middle term in the denominator of the transfer function (3.2-24) is

$$\lim_{\lambda \to \infty} \frac{A\rho_w c_w s}{\sqrt{\frac{\rho_w c_w}{\lambda}s}\, mc_f}\, th(\sqrt{\frac{\rho_w c_w}{\lambda}s}\, L) = \frac{AL\rho_w c_w}{mc_f}s \quad . \qquad (3.2\text{-}26)$$

Referring to (3.2-26), (3.2-24) becomes

$$G(s) = \frac{1}{\frac{M}{m}s + \frac{M_w c_w}{mc_f}s + 1} = \frac{1}{\frac{Mc_f + M_w c_w}{mc_f}s + 1} \qquad (3.2\text{-}27)$$

Thus G(s) in Equation (2.3-40) is the limiting case of Equation (3.2-24) for $\lambda \to \infty$, and equals the above Equation (3.2-27).

The next step in the analysis of heat conduction dynamics would be to solve PDE (3.2-5) for some given geometry of the body (slab, cylinder, semi-infinite solid bodies, sphere and so on) and given initial and boundary conditions. Such an analysis would surpass the planned framework and purpose of this book. It is the subject of whole chapters and books on thermodynamics and the equations of mathematical physics that treat the more limited problems of unsteady heat conduction, diffusion processes and the like, and give very detailed solutions and methods for solving parabolic PDE. Nevertheless, this book does cover an analysis of the dynamics of heat conduction through pipe walls of the type found in classical heat exchangers, in evaporator surfaces of steam generator and the like.

A common feature of most pipe exchangers is that the wall thickness is much smaller than the pipe radius (which makes it possible to neglect surface curvature) so that heat conduction through the wall can be observed as heat conduction through a flat surface.

The following example will show how to obtain a transfer function, how to analyze the dynamic properties and related characteristics (poles and zeros of transfer functions) of heat conduction processes through the pipe of the heat exchanger, and especially show how to reduce the order of the system and obtain transfer functions in the form of proper rational function.

Example 3 Heat conduction through the exchanger wall

For the pipe wall of the heat exchanger shown on Figure 3.2-3 determine transfer functions describing changes in the heat flow rate $q_2(t)$ transferred from the wall onto the heated fluid depending on heat flow rate $q_1(t)$ brought to the pipe by heating gases and the temperature $\vartheta_f(t)$ of the heated fluid. The dynamics of three cases will be analyzed:

a)

    - forced radiation heat flow rate $q_1(t)$ on the side of the heating
gases

    - very high coefficient of convective heat transfer α_2 onto the
heated fluid

b)

    - forced radiation heat flow rate $q_1(t)$ on the side of the heating
gases

    - finite coefficient of convective heat transfer α_2 onto the
heated fluid

c)

    - convective heat transfer on both surfaces of the wall with
finite heat transfer coefficients α_1 and α_2.

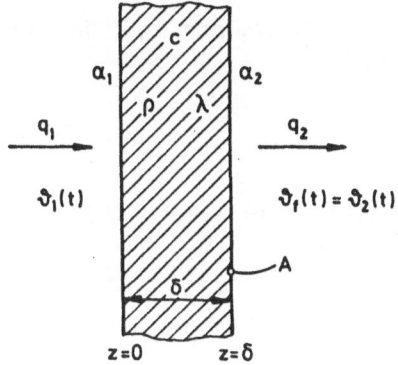

Fig. 3.2-3 Heat transfer through exchanger wall

The conditions a) and b) correspond to conditions that are often
realized in steam generator evaporator pipes.

a) Forced heat flow rate $q_1(t)$ and an infinite heat transfer coefficient α_2
get the following mathematical form for boundary conditions in $z = 0$
and $z = \delta$

1st BC $-\lambda A \dfrac{\partial \vartheta(0,\ t)}{\partial z} = q_1(t)$, $z = 0$.

2nd BC $\vartheta(\delta, t) = \vartheta_f(t)$, $z = \delta$.

Substituting the BC in cross-section $z = 0$ into Equation (3.2-10), which has first been differentiated with respect to z, and the BC for $z = \delta$ into Equation (3.2-10), yields constants C_1 and C_2

$$C_1 = \frac{\Theta_f(s) + \dfrac{e^{\sqrt{\frac{c\rho}{\lambda}s}}}{\sqrt{\frac{c\rho}{\lambda}s}}\dfrac{Q_1(s)}{\lambda A}}{2\,ch(\sqrt{\frac{c\rho}{\lambda}s}\,\delta)} \qquad . \tag{3.2-28}$$

$$C_2 = C_1 - \frac{Q_1(s)}{\lambda A\sqrt{\frac{c\rho}{\lambda}s}} \qquad . \tag{3.2-29}$$

The final expression for temperature distribution along the z axis in the wall is obtained after the above expressions are inserted into (3.2-10).

The following equation determines heat flow rate from the wall onto the heated fluid in the s-domain

$$Q_2(s) = -\lambda A\frac{d\Theta(\delta, s)}{dz} \qquad . \tag{3.2-30}$$

After $z = \delta$ is inserted into $d\Theta(z, s)/dz$, multiplication by $-\lambda A$ and rearrangement yields

$$Q_2(s) = \underbrace{\frac{1}{ch(\sqrt{\frac{c\rho}{\lambda}s}\,\delta)}}_{G_1(s)} Q_1(s) - \underbrace{\lambda A\sqrt{\frac{c\rho}{\lambda}s}\,\,th(\sqrt{\frac{c\rho}{\lambda}s}\,\delta)\,\,\Theta_f(s)}_{G_2(s)} \tag{3.2-31}$$

The transfer function $G_1(s)$ describes the dynamics of heat conduction through the pipe wall, and $G_2(s)$ the dynamics of heat flow rate change resulting from temperature changes of the heated fluid.

To investigate the dynamic properties of heat conduction we must analyze the denominator of the transfer function $G_1(s)$. The term $c\rho\delta^2/\lambda$ has the dimension of time, so we will represent it as the time constant T_w (it is useful to compare T_w with the time constant from Equation (2.3-14)).

$$T_W = \frac{c\rho\delta^2}{\lambda} \quad . \tag{3.2-32}$$

The denominator of $G_1(s)$ now becomes $ch\sqrt{T_W}\, s$. For a finite s this function has no singularities, so the roots of

$$ch\sqrt{T_W s} = 0 \quad . \tag{3.2-33}$$

are the poles of the transfer function $G_1(s)$. Introducing the substitution

$$\sqrt{T_W s} = x + jy \quad . \tag{3.2-34}$$

Equation (3.2-33) becomes

$$chx \; cosy + jshx \; siny = 0 \quad . \tag{3.2-35}$$

This is satisfied only when both the real and imaginary parts equal zero, which is fulfilled for $x = 0$ and for all y that satisfy

$$cosy = 0 \quad . \tag{3.2-36}$$

i.e.

$$y_k = (2k - 1)\frac{\pi}{2} \quad , \quad k = 0, \pm1, \pm2, \ldots \tag{3.2-37}$$

Inserting these values into (3.2-34) yields the expression for the poles of the transfer function $G_1(s)$

$$s_k = -\frac{(2k - 1)^2\pi^2}{4} \frac{1}{T_W} \quad , \quad k = 1, 2, 3, \ldots \tag{3.2-38}$$

(Since for $k < 0$ there is $s_0 = s_1$ and $s_k = s_k + 1$, it is enough to take only positive values for k, which is sufficient to determine all the poles.)

The expression (3.2-38) indicates that $G_1(s)$ has an **unlimited number of negative, real and different poles**. The fact that the poles have no imaginary parts shows that the process of heat conduction (and this will also be true of all other processes with equalization described by a parabolic PDE (3.2-1)) cannot have an oscillatory (periodic) character.

When the poles have been determined, the transfer function can be written in the form of an infinite product

$$G_1(s) = \frac{1}{ch\sqrt{T_w s}} = \prod_{k=1}^{\infty} \frac{s_k}{(s - s_k)} = \frac{s_1}{(s-s_1)} \frac{s_2}{(s-s_2)} \frac{s_3}{(s-s_3)} \ldots \qquad (3.2\text{-}39)$$

In other words, $G_1(s)$ can be replaced by an unlimited number of proportional first-order systems in series with **different time constants**, which can be calculated from

$$T_k = - \frac{1}{s_k} \quad , k = 1, 2, 3, \ldots \qquad (3.2\text{-}40)$$

Table 3.2-1 for the case of a steel evaporator pipe of wall thickness 5 mm ($\lambda = 46.5$ W/mK, c = 500 J/kgK, $\rho = 7850$ kg/m^3) gives the values for the first five poles and their corresponding time constants.

Table 3.2-1

k	s_k	T_k [s]
1	-1,169	0,855
2	-10,523	0,095
3	-29,232	0,034
4	-57,295	0,018
5	-94,714	0,011

Figure 3.2-1 shows the distribution of the poles in the s-plane.

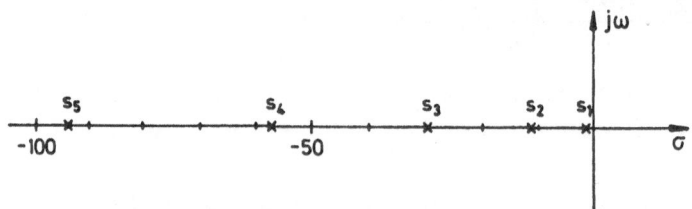

Fig. 3.2-4 Position of transfer function poles for processes with equalization

Here it is very important to observe the gradient of decrease of time constants (and this is not a function of the thickness or physical properties of the wall) where the 2nd, 3rd, 4th ... time constants are about 10, 30, 60 and 100 times smaller than the first. They decrease $\pi^2(2k-1)^2/4$ times. This distribution of time constants shows that it is possible to replace the transcendental transfer function $G_1(s)$ with a rational transfer function of a proportional first-order system whose time constant is similar to the first, dominant time constant in Equation (3.2-39). In this way we would, however, neglect all the very rapid transient processes which are characterized by the 2nd and higher time constants in the product (3.2-39). This was already done in the section on lumped parameters (Equation (2.3-14)), and here it remains for us to show the kind of thinking that made it possible.

Besides being expanded into the product shown, $G_1(s)$ can mathematically also be expanded as follows

$$G_1(s) = \frac{1}{ch\sqrt{T_w s}} = \frac{1}{1 + \frac{(\sqrt{T_w s})^2}{2!} + \frac{(\sqrt{T_w s})^4}{4!} + \dots} \qquad (3.2-41)$$

If all the terms of 4th and higher order are neglected, rearrangement of the upper equation yields the approximated transfer function $G_{1a}(s)$ which has already appeared in (2.3-14)

$$G_{1a}(s) = \frac{1}{\frac{c\rho\delta^2}{2\lambda}s + 1} \qquad (3.2-42)$$

The time constant characterizing Equation (3.2-42) for the above defined pipe wall is 1.055 s, which is only slightly more than the value of $T_1 = 0.855$ s.

Finally, we will give a graphical representation of changes in heat flow rate $q_2(t)$ if the flow rate $q_1(t)$ of the heating gases suddenly increases by a unit value. Figure 3.2-5 shows two curves: curve a shows the real changes of $q_2(t)$ given by the transfer function (3.2-39), and curve b shows the response $q_2(t)$ given by the approximate model, Equation (3.2-42).

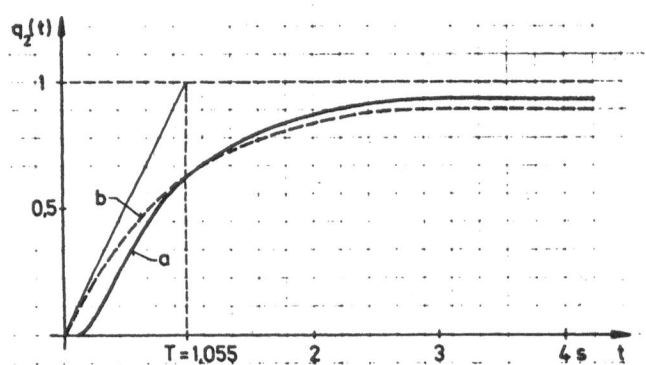

Fig. 3.2-5 Change in heat flow rate $q_2(t)$ in the case of a step increase in $q_1(t)$ on the heater side

The curves show that there are differences at the beginning of the transient process (for small t) that result from neglecting poles that are more distant in the negative direction of the real axis, i.e. because fast, high-frequency response components have been neglected. Of course, the newly achieved steady state will be the same in both cases, and for t →∞, $q_2(t)$→$q_1(t)$.

The transfer function $G_2(s)$ shows how the heat flow rate that is transferred from the pipe wall onto the evaporating fluid is affected by temperature change $\vartheta_f(t)$ of the fluid (in the evaporator this is caused by changes in the pressure of evaporation). To analyze the dynamic properties of that process further, $G_2(s)$ can be shown as follows

$$G_2(s) = \frac{Q_2(s)}{\theta_f(s)} = -\lambda A \sqrt{\frac{c\rho}{\lambda}} s \frac{sh(\sqrt{\frac{c\rho}{\lambda}} s \, \delta)}{ch(\sqrt{\frac{c\rho}{\lambda}} s \, \delta)} \qquad . \qquad (3.2\text{-}43)$$

The denominator of this transfer function equals that of $G_1(s)$, so the poles will be the same as the ones that have just been derived and expressed by Equation (3.2-38). $G_2(s)$, however, also has zeros and they influence the transient process as well. They can be obtained from equation

$$\lambda A \sqrt{\frac{c\rho}{\lambda}} s \; sh(\sqrt{\frac{c\rho}{\lambda}} s \, \delta) = 0 \qquad . \qquad (3.2\text{-}44)$$

If $K_p = \lambda A/\delta$ W/K is introduced and the above equation divided by this K_p, referring to (3.2-32), Equation (3.2-44) yields

$$\sqrt{T_w s}\ sh\sqrt{T_w s} = 0 \quad . \tag{3.2-45}$$

The substitution given by Equation (3.2-34) is also used here, so the last equation becomes

$$(x + jy)(shx\ cosy + jchx\ siny) = 0 \quad . \tag{3.2-46}$$

This product equals zero if any of the factors equal zero, whence the demand $x = 0$ and $siny = 0$. The final expression for the zeros of the transfer function $G_2(s)$ is

$$n_k = - \frac{k^2\pi^2}{T_w}\ , \quad k = 0,\ 1,\ 2,\ ... \tag{3.2-47}$$

Figure 3.2-6 shows the first 5 zeros and poles of the transfer function $G_2(s)$ in the s-plane. For the given steel pipe from (3.2-47) we get: $n_1 = 0$, $n_2 = -4.677$, $n_3 = -18.71$, $n_4 = -42.1$, $n_5 = -74.83$.

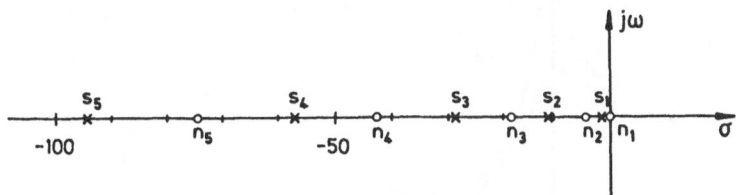

Fig. 3.2-6 Zeros and poles of transfer function $G_2(s) = Q_2(s)/\theta_f(s)$

On the basis of the analysis and procedure analogous to the one we used to obtain Equation (3.2-39), the transfer function $G_2(s)$ can be written in the form of the following infinite product

$$G_2(s) = - K_p \frac{\sqrt{T_w s}\ sh\sqrt{T_w s}}{ch\sqrt{T_w s}} = - K_p \prod_{k=1}^{\infty} \frac{s(s-n_k)}{(s-s_k)} = \tag{3.2-48}$$

$$= - K_p\ \frac{s(s-n_1)(s-n_2)...(s-n_n)...}{(s-s_1)(s-s_2)...(s-s_n)...} \quad .$$

Two essential facts are obvious: the heat flow rate $q_2(t)$ shows derivative dependence on fluid temperature change $\vartheta_f(t)$ (the appearance of an independent s in the numerator of $G_2(s)$), and since all the poles are negative and real, $q_2(t)$ has no oscillatory properties in this case also.

If we want to show how $q_2(t)$ changes with time if $\vartheta_f(t)$ undergoes a step unit **decrease**, the following equation gives an exact analytical solution

$$q_2(t) = - K_p \mathcal{L}^{-1} \left\{ \frac{1}{s} \frac{\sqrt{T_w s} \ sh\sqrt{T_w s}}{ch\sqrt{T_w s}} \right\} \ . \qquad (3.2\text{-}49)$$

After inverse Laplace transformation we get $q_2(t)$ in the form of an infinite sum of exponential functions

$$q_2(t) = K_p \sqrt{\frac{T_w}{\pi t}} \ (1 + 2 \sum_{k=1}^{\infty} (-1)^k \ e^{-k^2 \frac{T_w}{t}}) \ . \qquad (3.2\text{-}50)$$

Investigation into $q_2(t)$ has shown that for $t = 0$, $q_2(t)$ is infinite and as t approaches infinity the density of heat flow rate $q_2(t)$ converges to zero. On Figure 3.2-7 curve a shows changes of the transfer function $q_2(t)$.

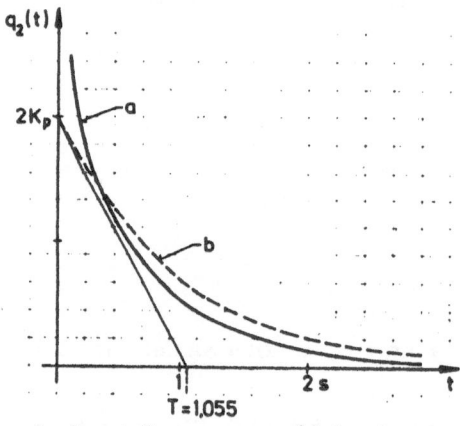

Fig. 3.2-7 Change in heat flow rate $q_2(t)$ in the case of a unit step temperature $\vartheta_f(t)$ decrease

Just after ($t^+ = 0$) the step decrease of ϑ_f heat energy stored in the surface layer of the wall is immediately released. This amount, however, is infinitesimally small (the heat flow rate $q_2(t)$ must not be confused with the heat released $Q_2(t)$) and as all the other layers of the wall in the z direction offer thermal resistance, the heat flow rate $q_2(t)$ decreases very quickly. For small t the gradients of change q_2 are very great, and only after the deep layers of the wall have been included in the process of heat conduction and transfer towards the fluid of lower temperature ϑ_f (which takes a certain finite period of time) does the change $q_2(t)$ become less steep.

Here again the complexity of $G_2(s)$ invites us to try to approximate this transcendental transfer function by a simpler function in the form of rational polynomials. The simplest approximation obtained from Equation (3.2-48) is to retain in the expansion of the function $G_2(s)$ only processes with the greatest time constant, i.e. to neglect all terms of a third and higher order in the expansion of transcendental hyperbolic functions. The approximate function $G_{2a}(s)$ is obtained from the following expansion

$$G_{2a}(s) = -K_p \frac{\sqrt{T_w s}\,(\sqrt{T_w s} + \cdots)}{1 + \frac{(\sqrt{T_w s})^2}{2!} + \cdots} = -K_p T_w \frac{s}{\frac{c\rho\delta^2}{2\lambda}s + 1} \qquad (3.2\text{-}51)$$

This has given us the well-known transfer function of the derivative system with first-order lag, whose response to unit step decrease in the temperature of the heated fluid is shown by curve b on Figure 3.2-7. As in Figure 3.2-5, here also the same remark holds that neglecting fast processes (higher-order terms of the development) leads to different responses at the beginning of the transient process, but that this difference between real and approximation-generated response curves decreases with the increase of time t.

This ends the analysis of the case when the coefficient of convective heat transfer on the side of the heated fluid can be considered infinite. The closest to this are evaporation processes in steam generator pipes. There the coefficients α_2 are really high, but not infinite, so it is of interest to examine the case (given by b) in this example) when the coefficient of convective heat transfer α_2 is finite.

b) Forced heat flow rate $q_1(t)$ and a finite coefficient of heat transfer α_2 lead to the following mathematical formulation of boundary conditions

1st BC $\qquad -\lambda A \frac{\partial \vartheta(0,\,t)}{\partial z} = q_1(t)$, $\quad z = 0$

2nd BC $\qquad \frac{\partial \vartheta(\delta,\,t)}{\partial z} = -\frac{\alpha_2}{\lambda}\,(\vartheta(\delta,\,t) - \vartheta_f(t)),\quad z = \delta$.

Like in the preceding case, C_1 and C_2 are determined depending on the boundary conditions

$$C_1 = \frac{\frac{\alpha_2}{\lambda}\Theta_f(s) + \frac{D(s)}{E(s)}\frac{Q_1(s)}{\lambda A}}{B(s) + D(s)} \quad . \tag{3.2-52}$$

$$C_2 = \frac{\frac{\alpha_2}{\lambda}\Theta_f(s) - \frac{B(s)}{E(s)}\frac{Q_1(s)}{\lambda A}}{B(s) + D(s)} \quad . \tag{3.2-53}$$

$$B(s) = (-\sqrt{\frac{C\rho}{\lambda}s} + \frac{\alpha_2}{\lambda})\, e^{-\sqrt{T_w s}} \quad , \tag{3.2-54}$$

$$D(s) = (\sqrt{\frac{C\rho}{\lambda}s} + \frac{\alpha_2}{\lambda})\, e^{\sqrt{T_w s}} \quad , \tag{3.2-55}$$

$$E(s) = \sqrt{\frac{C\rho}{\lambda}s} \quad . \tag{3.2-56}$$

The temperature of the pipe wall along the z axis $\Theta(z, s)$ is determined by Equation (3.2-10), and $z = \delta$ gives the temperature on the surface facing the heated fluid. If we know $\Theta(\delta, s)$, the heat flow towards the fluid is determined by

$$Q_2(s) = Q(\delta, s) = \alpha_2 A\left[\Theta(\delta, s) - \Theta_f(s)\right] \quad . \tag{3.2-57}$$

Substituting $\Theta(\delta, s)$ into this equation and arranging it yields the expression demanded

$$Q_2(s) = \underbrace{\frac{1}{ch\sqrt{T_w s} + \frac{\lambda}{\alpha_2\delta}\sqrt{T_w s}\, sh\sqrt{T_w s}}}_{G_3(s)} Q_1(s) -$$

$$- K_p\underbrace{\frac{\sqrt{T_w s}\, sh\sqrt{T_w s}}{ch\sqrt{T_w s} + \frac{\lambda}{\alpha_2\delta}\sqrt{T_w s}\, sh\sqrt{T_w s}}}_{G_4(s)}\Theta_f(s) \quad . \tag{3.2-58}$$

In the further text we will not repeat the methods and procedures used in the preceding case, especially since these calculations are much more complex and the results obtained do not show any essentially new quality. The basic characteristics of the process for the above boundary conditions are the following.

The poles of the transfer functions $G_3(s)$ and $G_4(s)$ are still negative and real and there is an infinite number of them, but now they cannot be given in explicit mathematical form. They are, therefore, obtained graphically-analytically and given in tabular form. The first pole, which also characterizes the process with the greatest time constant, is determined by the following approximate formula for the case of $(\alpha_2 \delta / \lambda) > 0.1$.

$$s_1 = -\frac{\pi^2}{4} \frac{1}{1 + 2.24(\frac{\alpha_2 \delta}{\lambda})^{-1.02}} \frac{1}{T_w} \tag{3.2-59}$$

The responses $q_2(t)$ to step changes $q_1(t)$ and $\vartheta_f(t)$ are still of unperiodic character and are completely similar in form to the preceding case, and the transfer function $q_2(t)$ again shows a derivative response in the case of disturbance in $\vartheta_f(t)$. Unlike the preceding case, however, $q_2(t)$ is no longer infinite for $t = t_0$.

It is interesting that very good approximations of transfer functions $G_3(s)$ and $G_4(s)$ can also be realized now if all the terms of third and higher order in the expansion of the transcendental functions $ch\sqrt{T_w s}$ and $sh\sqrt{T_w s}$ are neglected. The responses obtained from these transformed functions are close to the real responses, but because higher-order terms have been neglected, which means all the fast, high-frequency parts of the responses that participate in the dynamics of the process, there are insignificant differences at the beginning of the transient process.

The approximated transfer function $G_{3a}(s)$ is obtained as follows

$$G_{3a}(s) = \frac{1}{(1 + \frac{(\sqrt{T_w s})^2}{2}) + \frac{\lambda}{\alpha_2 \delta} \sqrt{T_w s} \sqrt{T_w s}} = \tag{3.2-60}$$

$$= \frac{1}{c\rho\delta(\frac{1}{\alpha_2} + \frac{\delta}{2\lambda})s + 1} .$$

This is obviously a transfer function of the proportional system of first order and it is important to note that the time constant obtained is in fact equal to the expanded (with the term $1/\alpha_2$) time constant that characterized heat conduction when it was observed as a process with lumped parameters: Equations (2.3-14) and (3.2-42).

If the hyperbolic functions comprising the transfer function $G_4(s)$ are expanded into a series and the higher-order terms neglected, $G_{4a}(s)$ is obtained

$$G_{4a}(s) = - K_p T_w \frac{s}{c\rho\delta(\frac{1}{\alpha_2} + \frac{\delta}{2\lambda})s + 1} \qquad (3.2-61)$$

The transfer functions $G_{3a}(s)$ and $G_{4a}(s)$ are completely identical with the approximated functions $G_{1a}(s)$ and $G_{2a}(s)$ shown by Equations (3.2-42) and (3.2-51), differing only in the time constant. It is important to note that $G_{1a}(s)$ and $G_{2a}(s)$ are special boundary cases of Equations (3.2-60) and (3.2-61) with the assumption of an infinite α_2.

Keeping to our usual (inductive) manner of presenting and analyzing dynamic properties and proceeding from the special and simple to the general and more complex, it still remains to analyze the case when **there in convective heat transfer with a finite value for the coefficient** α_1 **in** $z = 0$ **also**. Here we will leave out completely the detailed procedures of obtaining transcendental transfer functions and only show their approximations.

c) Finite coefficients of heat transfer α_1 and α_2

In this case of convective heat transfer with finite transfer coefficients in $z = 0$ (α_1) and $z = \delta$ (α_2), the boundary conditions become equal to those given in c) of Example 1.

1st BC $\qquad \frac{\partial\vartheta(0, t)}{\partial z} = \frac{\alpha_1}{\lambda}(\vartheta(0, t) - \vartheta_1(t))$, $z = 0$

2nd BC $\qquad \frac{\partial\vartheta(\delta, t)}{\partial z} = - \frac{\alpha_2}{\lambda}(\vartheta(\delta, t) - \vartheta_2(t))$, $z = \delta$

$\vartheta_1(t)$ and $\vartheta_2(t)$ are now the temperatures of the media flowing on the outside and the inside of the wall where $\vartheta_1 > \vartheta_2$.

The approximated transfer functions in Equation (3.2-62) show how heat flow rate $q_2(t)$, transferred onto the heated fluid, depends on temperature changes of both the heating and the heated fluid.

$$Q_2(s) = \underbrace{\frac{K}{T_n s + 1}}_{G_{5a}(s)}\Theta_1(s) - \underbrace{\frac{K(T_b s + 1)}{T_n s + 1}}_{G_{6a}(s)}\Theta_2(s) \quad . \tag{3.2-62}$$

$$K = \frac{\alpha_1 \alpha_2}{\alpha_1 + \alpha_2 + \alpha_1 \alpha_2 \dfrac{\delta}{\lambda}} A \tag{3.2-63}$$

$$T_b = c\rho\delta\left(\frac{1}{\alpha_1} + \frac{\delta}{2\lambda}\right) \tag{3.2-64}$$

$$T_n = c\rho\delta \frac{1 + (\alpha_1 + \alpha_2)\dfrac{\delta}{2\lambda} + \alpha_1 \alpha_2 \dfrac{\delta^2}{6\lambda^2}}{\alpha_1 + \alpha_2 + \alpha_1 \alpha_2 \dfrac{\delta}{\lambda}} \quad . \tag{3.2-65}$$

In the above equations we have assumed that the surface areas through which heat is transferred are the same on the outer and on the inner side of the wall, i.e. we worked with a mean surface area $A = A_1 = A_2$.

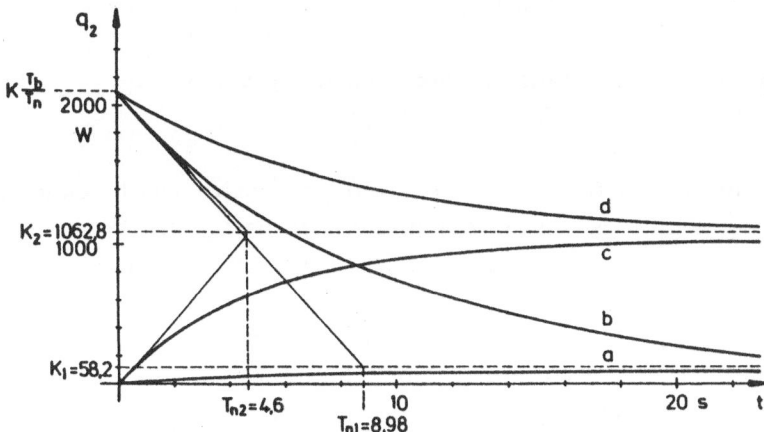

Fig. 3.2-8 Heat flow rate change $q_2(t)$ in the case of:
- step temperature increase $\vartheta_1(t)$ - a,c
- step temperature decrease $\vartheta_2(t)$ - b,d

The transfer functions $G_{5a}(s)$ and $G_{6a}(s)$ show that $q_2(t)$ will respond as a proportional first-order system to a disturbance on the heating side (a disturbance that passes through the pipe wall). To temperature disturbances of the heated medium it will continue to show derivative properties, only now $q_2(t)$ will no longer converge to zero after a disturbance of $\vartheta_2(t)$. This property fundamentally distinguishes this transient process from the response in the preceding two cases when there was a forced heat flow rate $q_1(t)$ in $z = 0$, when $q_1(t)$ was not a function of the wall temperature $\vartheta(0, t)$, i.e. when there was no feedback action of the temperature of the heated medium $\vartheta_2(t)$ on the amount of heat brought to the pipe wall (and through it to the heated fluid).

The transfer function $G_{6a}(s)$ has qualitatively different transfer properties so we will show its responses in this case when the coefficients of heat transfer on both sides of the pipe are finite. For steam generator superheater pipes whose data have already been given in the section on lumped parameters, Figure 3.2-8 shows the responses obtained on the basis of Equation (3.2-62) in the case of unit step temperature increase $\vartheta_1(t)$ (curve a) and decrease $\vartheta_2(t)$ (curve b).

The following data are valid for the metal wall:

c = 500 J/kgK, ρ = 7850 kg/m^3, δ = 0.005 m, λ = 46.5 W/mK, α_1 = 60 W/m^{2}K, α_2 =2400 W/m^{2}K, A = 1 m^2 .

The coefficients characterizing the transfer process are K = 58.1704 W/K, T_b = 328.138 s, T_n = 8.98 s.

It is obvious that heat flow rate $q_2(t)$ increases much more for a decrease in ϑ_2, which is the consequence of using up heat that was stored (accumulated) in the pipe wall. When ϑ_1 increases not only is no stored heat released from the wall, but the greater heat flow is used to increase the energy level of the wall itself, which results in a much smaller growth of $q_2(t)$. Nevertheless, this difference in the value of $q_2(t)$ results from the much greater heat resistance in the cross-section $z = 0$ (α_1 is about 40 times smaller than α_2). Thus, Figure 3.2-8 also shows the transfer processes if the coefficient α_1 is much greater than above and equal to α_2, α_1 = α_2 = 2400 W/m^{2}K. Then K = 1062.8 W/K, T_b = 9.232 s, and T_n = 4.6 s. Curve c corresponds to temperature increase $\vartheta_1(t)$, curve d to temperature decrease $\vartheta_2(t)$.

To end this part we must still show that the analysis in Section 2.3

gave time constants that are boundary cases of Equations (3.2-63), (3.2-64) and (3.2-65) for an infinite λ. If these time constants are compared it can be seen that the differences are not very great. However, it must be repeated that if the coefficients of heat transfer α_1 and α_2 are large, which means that the convective part of the heat resistance is small, then the part of the resistance resulting from heat conduction through the wall makes up a large part of the total time constant T and thus Equations (2.3-10), (2.3-11) and (2.3-12) are not satisfactory.

In all the previous examples we used the basic form of the parabolic PDE shown by Equation (3.2-1) to analyze the dynamic properties of processes with equalization, and any differences and difficulties we encountered in solving it resulted from the necessity of satisfying different conditions of heat transfer on the boundaries of the system under observation. The preceding pages show the complexity of the mathematical tools and the necessity of turning to simpler forms of transfer functions. All the solutions that we obtained here were solutions of the simplest forms of parabolic PDE, but it is several times more difficult to get solutions analytically if the process occurs in a body with a more complicated geometry, or if a more "extended" form of Equation (3.2-1) describes the process. In such cases a digital computer and numerical methods of solution must be used. The following two examples will show how a very small change in assumptions (whose existence and help are crucial when the equations are formulated) results in a different (more complex) mathematical form of the basic parabolic PDE (3.2-1).

Example 4 Heat conduction through homogeneous uninsulated body

Derive an equation describing unsteady temperature changes for the body on Figure 3.2-1 if all its side surfaces exchange heat with an environment of temperature $\vartheta_0(t)$. Resistance to heat conduction in a cross-section is neglected and the model is formulated for the case when the surrounding temperature is greater than the temperature of the body.

If the circumference of the body is U and if α is the coefficient of heat transfer on the side-surface boundary of the body, the left-hand side of the equation for the conservation of energy (3.2-2) is expanded by terms containing the amount of heat brought to volume dV through the side surfaces

$$q(z, t)A - \left[q(z, t) + \frac{\partial q(z, t)}{\partial z} dz \right] A - \alpha U \left[\vartheta(z, t) - \vartheta_0(t) \right] dz =$$

$$= A \rho c \frac{\partial \vartheta(z, t)}{\partial t} dz \qquad (3.2\text{-}66)$$

Arranging this upper equation gives the final PDE

$$\frac{\partial \vartheta(z, t)}{\partial t} - \frac{\lambda}{c\rho} \frac{\partial^2 \vartheta(z, t)}{\partial z^2} + \frac{\alpha U}{Ac\rho} \vartheta(z, t) = \frac{\alpha U}{Ac\rho} \vartheta_0(t) \quad . \qquad (3.2\text{-}67)$$

If the coefficient α can be considered independent of temperature, a nonhomogeneous linear parabolic PDE is obtained. The boundary and initial conditions defined for Example 1 are completely valid for Equation (3.2-67) also, but to get a unique solution of the equation it is necessary to know the law of temperature change $\vartheta_0(t)$ in the environment as well.

Example 5 Process of diffusion with convection

Consider a pipe of cross-sectional area A through which a solution flows with velocity w. The solution contains a C component with varying concentration c along the pipe. Derive an equation describing dynamic changes along the pipe in the concentration of component C. Concentration c is considered constant on the cross-sectional area and Fick's law is valid for diffuse flow.

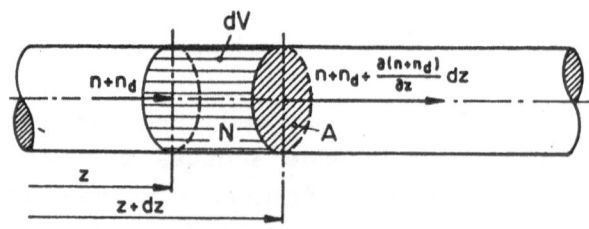

Fig. 3.2-9 Diffusion with convection

Figure 3.2-9 shows an elementary pipe volume dV for which the equation for mass conservation of the C component in the solution is to be derived. The following notations have been used: n ... C component molar flow transported by the flow of the solution (convective molar flow), $n = cAw$ mol C/s, c ... concentration of C component mol C/m^3, N ... amount of C component, $N = cV$ mol C, n_d ... diffuse molar flow (results from concentration difference along z), $n_d = -DA \, \partial c/\partial z$ mol C/s (Fick's law), D .... coefficient of diffusion m^2/s.

$$\left[n(z, t) + n_d(z, t) \right] - \left[n(z, t) + n_d(z, t) + \frac{\partial \left[n(z, t) + n_d(z, t) \right]}{\partial z} dz \right] = \frac{\partial N}{\partial t}$$

$$(3.2\text{-}68)$$

$$- \frac{\partial \left[n(z, t) + n_d(z, t) \right]}{\partial z} dz = \frac{\partial (c(z, t) dV)}{\partial t} = \frac{\partial c(z, t)}{\partial t} A dz \qquad (3.2\text{-}69)$$

$$- \frac{\partial c(z, t)}{\partial z} Aw - \frac{\partial \left[-DA \frac{\partial c(z, t)}{\partial z} \right]}{\partial z} = \frac{\partial c(z, t)}{\partial t} A \qquad (3.2\text{-}70)$$

Rearrangement of the last equation for a constant coefficient of diffusion D, which means it is not a function of c, yields

$$\frac{\partial c(z, t)}{\partial t} + w \frac{\partial c(z, t)}{\partial z} - D \frac{\partial^2 c(z, t)}{\partial z^2} = 0 \quad . \qquad (3.2\text{-}71)$$

This is a linear homogeneous parabolic PDE. Compared to the basic equation for processes with equalization (3.2-1), it is expanded by the second term on the left-hand side which contains the so-called convective part of the transport of component C along the z axis, the part that is carried with the solution flow by velocity w.

These last two examples complete the presentation of some basic characteristics of processes with equalization. The following section treats periodic processes whose basic mathematical notation (for processes with distributed parameters) is a second-order hyperbolic PDE. The major difference characterizing the dynamics of earlier processes in comparison with these processes is that the poles of the transfer functions will now be complex conjugate pairs. This distribution of poles is the result of properties of inertia possessed by continua, which appear in the momentum equation when the conservation equations are formulated. Such properties are not shown in processes with equalization so there was no need to formulate the equation for the conservation of momentum for them.

3.3 PROCESSES WITH PERIODIC STATE CHANGES
BASIC SECOND-ORDER
HYPERBOLIC PARTIAL DIFFERENTIAL EQUATION

Oscillations (vibrations) in parts of devices and plants are an everyday occurrence in technical practice and many different processes demonstrate these periodic dynamic properties. Vibrations of rods, beams, shafts, shells and water masses in hydro-electric power plant feeding pipes, water hammer effects in hydraulic and pneumatic pipes, oscillations of water level in connected tanks, oscillations in electrical circuits, torsional oscillations and so on are all examples of dynamically the same or similar phenomena whose state variables are related by the same type of differential equation. If processes whose parameters change along the spatial axis are analyzed, then the basic and simplest form of equation describing periodic state changes will be the following **second-order hyperbolic PDE**.

$$\frac{\partial^2 u}{\partial t^2} = a^2 \frac{\partial^2 u}{\partial z^2} \quad .$$

(3.3-1)

Equation (3.3-1) was derived after making many assumptions that excluded phenomena unimportant for the purposes of this analysis, and it is valid for different processes in which the variables u and a have the following physical meaning:

longitudinal oscillation of a beam

  u ... displacement of cross-section

  a ... velocity of disturbance propagation along the beam, $a = \sqrt{E/\rho}$

  E ... modulus of elasticity (Young's modulus)

  ρ ... density

hydraulic shocks in pipe

  u ... pressure, flow, velocity, density

  a ... speed of sound (speed of disturbance propagation)

oscillation in electrical circuits

        u ... voltage, current flow

        $a = \sqrt{LC}$, L ... circuit inductance

        C ... circuit capacitance

To obtain a unique solution of Equation (3.3-1) it is necessary to know the initial conditions (IC) and the boundary conditions (BC), whose mathematical formulation contains data about whether, and how much, energy, mass or momentum was stored in the process under observation at the initial moment (IC), and how those variables are exchanged with the environment through the boundaries (BC). In mathematical form the BC of the second-order hyperbolic PDE are equal or similar to the BC from the preceding section for the parabolic PDE. In the case of the IC, since Equation (3.3-1) now contains the second derivative of variable u with respect to time t, besides knowing function $u_0 = u(z, 0)$ it is also necessary to know the value of function $u'_0 = du(z, 0)/dt$. To solve the above equation it is, therefore, necessary to formulate **two** initial and **two** boundary conditions in accordance with the real conditions under which the process occurs. Special attention must be paid to the mathematical formulation of conditions under which the system communicates with the environment through its boundaries. In the case of one-dimensional spatial distribution (and here we will treat such cases) it is always necessary to define conditions on both the boundaries of the process, $z = 0$ and $z = L$. The problem cannot be solved if both boundary conditions are given on one boundary because then the way in which the process occurs at the other end of the system is not known. This is the simplest explanation of the statement made in the section on fluid flow regarded as a process with lumped parameters, where it was said that both flow change and pressure change cannot be simultaneously given as disturbance variables at the same end of the pipe.

In this part of the book, where we treat periodic changes of state, most of our attention will given to the analysis of dynamic processes that occur inside pipes for transporting fluid. Oscillations in electromagnetic and mechanical systems are treated in great detail in specialized books from these fields. Nevertheless, although the accent in the further lines will be on hydraulic and pneumatic processes, we will also try to indicate the dynamic characteristics common to all periodic processes independent of the system in which they occur. Therefore, we will begin with a classical example of periodic state changes (longitudinal vibration of a beam) and use it to show the variety of possible boundary conditions. Later, when we analyze hydraulic oscillations, the similarity of the dynamic properties with this case from mechanics will be seen.

Example 1 Longitudinal vibration of a beam

Figure 3.3-1 shows a prismatic body (beam, bar) which performs longitudinal vibrations due to the action of a force. Assuming that the beam is of constant cross-sectional area, homogeneous, that its bending is neglected and that longitudinal displacements occur within the field of elastic deformations, derive the model describing that periodic process and define boundary conditions if:

  a - the left end is fixed, the right end free
  b - force F acts on the left end, the right end is fixed
  c - the left end is fixed, the right end carries mass M.

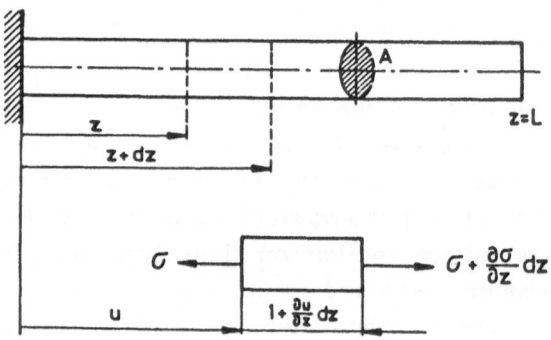

Fig. 3.3-1 Sketch of longitudinal beam vibration

The equation for the conservation of momentum, in this case more usual as the equation of force equilibrium, is formulated for an infinitesimal mass of the beam dM which is at the distance u in the stretched state and has the length $(1 + \partial u/\partial z)dz$. (In the steady state that same mass is at the distance z and is of length dz.)

The equation for force equilibrium is

$$- \sigma A + (\sigma + \frac{\partial \sigma}{\partial z} dz)A = \rho A dz \frac{\partial^2 u}{\partial t^2} \quad . \tag{3.3-2}$$

σ denotes tension in the cross-section, of dimension N/m^2. The above equation is reduced to

$$\frac{\partial \sigma}{\partial z} = \rho \frac{\partial^2 u}{\partial t^2} \quad . \tag{3.3-3}$$

In the region of elastic deformations there is

$$\sigma = E\varepsilon = E\frac{\partial u}{\partial z} \quad .$$

where ε denotes the linear deformation, and E the modulus of elasticity N/m^2. The last two equations yield

$$c^2 \frac{\partial^2 u}{\partial z^2} = \frac{\partial^2 u}{\partial t^2} \quad , \quad c = \sqrt{\frac{E}{\rho}} \quad . \tag{3.3-4}$$

The above equation has the already-known form of a second-order hyperbolic PDE that describes periodic processes. To solve it, it is necessary to know **two** boundary and **two** initial conditions. Here we will not enter into how to obtain solutions but will only give the **general form** for the solution of the above equation

$$u(z, t) = (C_1 \sin\omega t + C_2 \cos\omega t)(C_3 \sin\frac{\omega}{c}z + C_4 \cos\frac{\omega}{c}z) \quad . \tag{3.3-5}$$

The constants C_1, C_2, C_3 and C_4 are determined from the boundary and the initial conditions. To correspond with the conditions demanded at the beginning of this example, the boundary conditions can be varied:

a) The left end is fixed, the right end free.

> In this case there can be no displacement in $z = 0$, and in $z = L$, since that end is free, there can be no tension, so there is

1st BC $u(0, t) = 0$

2nd BC $\frac{\partial u(L, t)}{\partial z} = 0$

b) Force F acts on the left end, the right end is fixed.

In z = 0 a force of tension opposes force F, and the conditions
of force equilibrium yields

1st BC $F = A\sigma_0 = AE\dfrac{\partial u(0, t)}{\partial z}$.

2nd BC $u(L, t) = 0$.

c) The left end is fixed, the right carries mass M.

In this case the inertia of mass M in the cross-section z = L
causes tension along the whole cross-sectional area A so the
second BC is obtained from the condition of equilibrium
between the inertia of mass M and the force of tension in the
cross-section

1st BC $u(0, t) = 0$,

2nd BC $M\dfrac{\partial^2(L, t)}{\partial t^2} = - A\sigma_L = -AE\dfrac{\partial u(L, t)}{\partial z}$

This ends our presentation of the simplest example of periodic
processes in systems with distributed parameters. Of course, it is possible
to give many other different boundary conditions, to seek for the natural
frequencies of vibrations or for the solution of Equation (3.3-4) for given
conditions and the like. We will leave this to more specialized textbooks.
Our next example will be one-dimensional fluid flow, where the same or
similar (more complicated, in fact) equations to these in the above lines
will be obtained for the description of periodic processes. It is our desire
to show that different processes have the same dynamic properties, and
also to describe in more detail processes of disturbance propagation
along the fluid streamlines.

Example 2 Dynamic processes in fluid pipes. Equations of mass,
energy and momentum conservation

Consider a pipe of constant cross-sectional area A externally heated
by heat flow of density q J/ms per unit length, through which flows

m kg/s mass of fluid. The flow is homogeneous with the same velocities and thermodynamic properties per cross-section. The thermal capacity of the pipe wall is neglected. Derive the mathematical model describing unsteady state variable changes of that fluid.

Here also, analogously to the previous process, we will obtain the model by deriving conservation equations for the fluid in a control volume dV. However, unlike the previous process, here there will be mass, momentum and energy storage in the control volume dV at the same time, so it will be necessary to formulate all three conservation equations. After they have been arranged, reduced to a suitable form and the thermodynamic equations of fluid state referred to, a model in the desired form will be obtained. As until now, this will be a differential equation. We must, nevertheless, mention that when problems from fluid dynamics are analyzed the conservation equations are often given in the integral form, which is not the most suitable form for the purposes of this analysis. Therefore, such models are not derived in this book. This remark can be of use in a situation when the reader encounters literature in which the same laws are described using different mathematical tools.

EQUATION FOR THE CONSERVATION OF MASS

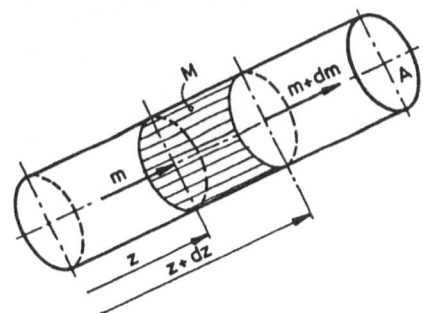

Fig. 3.3-2 Sketch for deriving continuity equation

If there are no sources or sinks of mass in dV, the difference between the amount of mass that enters dV and the amount that emerges from it in the time dt equals the change of the mass dM in dV. If m and m+dm show mass flow rates through cross-sections z and z+dz, then the amount of mass that enters dV during the time dt equals mdt, and the amount that emerges is (m+dm)dt. Thus we can write

$$\text{mdt} - (m + dm)dt = dM \tag{3.3-6}$$

In the generally unsteady state mass flow rate is variable both along the pipe and in time, i.e. m = m(z, t), so that flow rate change along the z coordinate, at time t, can be written

$$dm = \frac{\partial m}{\partial z} dz \quad . \tag{3.3-7}$$

The last two equations yield

$$- \frac{\partial m}{\partial z} dz = \frac{dM}{dt} = \frac{d(A\rho dz)}{dt} = Adz \frac{d\rho}{dt} \tag{3.3-8}$$

$$A\frac{\partial \rho}{\partial t} + \frac{\partial m}{\partial z} = 0 \quad . \tag{3.3-9}$$

Equation (3.3-9) is the law for the conservation of mass. However, as the same law is encountered in various forms in literature, it is advantageous to present those various forms of the continuity equation here also. It must be remembered that mass flow rate m is

$$m = A w \rho \quad . \tag{3.3-10}$$

and for $\rho = 1/v$ from (3.3-9), the following forms of the continuity equation can be obtained

$$\frac{\partial \rho}{\partial t} + w\frac{\partial \rho}{\partial z} = -\rho\frac{\partial w}{\partial z} \tag{3.3-11}$$

$$\frac{\partial v}{\partial t} + w\frac{\partial v}{\partial z} = v\frac{\partial w}{\partial z} \tag{3.3-12}$$

$$\frac{\partial m}{\partial t} + w\frac{\partial m}{\partial z} = A\rho\frac{\partial w}{\partial t} = \frac{m}{w}\frac{\partial w}{\partial t} \tag{3.3-13}$$

The last three expressions contain a derivative that is characteristic of fluid flow processes. It is called the material or substantial derivative and denoted D/dt

$$\frac{Df}{dt} = \frac{\partial f}{\partial t} + w\frac{\partial f}{\partial z} \tag{3.3-14}$$

The material or substantial derivative D/dt will appear in all conservation equations that describe state changes of flowing fluid. The name itself comes from the fact that every change depends both on the

local gradient of change $\partial/\partial t$, and on the so-called **convective** gradient of change, which depends on the velocity of "material (substance)" flow $w\partial/\partial z$.

Using (3.3-14), we can write

$$\frac{D\rho}{dt} = -\rho\frac{\partial w}{\partial z} \quad , \quad \frac{Dv}{dt} = v\frac{\partial w}{\partial z} \quad , \quad \frac{Dm}{dt} = \frac{m}{w}\frac{\partial w}{\partial z}$$

In steady state, in which there must be $\partial/\partial t = 0$, Equations (3.3-9) - (3.3-13) yield the following equalities

$$\overline{m}_z = \overline{m}_{z+dz} = \overline{m} \quad , \tag{3.3-15}$$

$$\overline{\rho w} = \text{const.} \quad . \tag{3.3-16}$$

For incompressible fluids (sometimes liquids can be considered incompressible) $\rho = \text{const.}$, and (3.3-16) yields

$$\overline{w}_z = \overline{w}_{z+dz} = \overline{w} = \text{const.} \quad . \tag{3.3-17}$$

EQUATION FOR THE CONSERVATION OF MOMENTUM

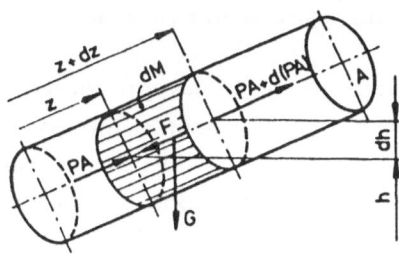

Fig. 3.3-3 Sketch for deriving momentum equation

The law for the conservation of momentum says that the sum of all forces acting on the mass of fluid dM in the direction of flow (the direction of the z axis) equal the change of momentum (dMw) of the particle dM in time dt.

Mathematically

$$\sum_{i=1}^{4} F_i = \frac{d(dMw)}{dt} = dM\frac{dw}{dt} \tag{3.3-18}$$

It must be observed that, unlike preceding case when the balance of mass was formulated for an elementary and unchangeable control volume dV, here the balance of momentum is formulated for an elementary and unchangeable control mass dM (which is in dV at the moment t). The fact that the equation of equilibrium is formed for an unchangeable (i.e. constant and independent of both z and t) particle of mass dM makes it possible to place the symbol dM in front of the operator of total differentiation d/dt in Equation (3.3-18).

The following forces act in the direction of the z axis:

pressure force in the cross-section z: $F_1 = PA$,

pressure force in the cross-section z+dz: $F_2 = PA + \frac{\partial(PA)}{\partial z}dz$,

component of gravitational force in the z direction:

$$F_3 = Gcos\alpha = dMg\frac{\partial h}{\partial z} =$$

$$= Adz\rho g\frac{\partial h}{\partial z} ,$$

friction force of fluid against the pipe wall: $F_4 = F = fAdz$.

In the above equations g is the gravitational acceleration and f the coefficient of friction, i.e. it represents the pressure drop due to friction ΔP_F per unit length of pipe, of the dimension Pa/m. Since force and momentum are vectors, their directions are important and if the increasing z direction is given with the + sign, (3.3-18) becomes

$$F_1 - F_2 - F_3 - F_4 = A\rho dz\frac{dw}{dt}$$

or

$$-\frac{\partial P}{\partial z} - \rho g\frac{\partial h}{\partial z} - f = \rho(\frac{\partial w}{\partial t} + w\frac{\partial w}{\partial z}) . \tag{3.3-19}$$

The total change dw is

$$dw = \frac{\partial w}{\partial t}dt + \frac{\partial w}{\partial z}dz = \left(\frac{\partial w}{\partial t} + \frac{\partial w}{\partial z}\frac{dz}{dt}\right)dt \qquad (3.3\text{-}20)$$

If fluid flows at velocity w, in period of time dt its particle dM will move along path dz = wdt, so quotient dz/dt is flow velocity w. This means that the expression in brackets in the above equation can be written in the form of material derivative Dw/dt and Equation (3.3-19), referring to (3.3-20), after rearrangement becomes

$$\rho\frac{Dw}{dt} + \frac{\partial P}{\partial z} + \underbrace{g\rho\frac{\partial h}{\partial z}}_{\Delta P_G} + \underbrace{f}_{\Delta P_F} = 0 \qquad (3.3\text{-}21)$$

ΔP_G ... gravitational pressure change **per unit length** Pa/m

ΔP_F ... frictional pressure change (drop) **per unit length** Pa/m

The equation for the conservation of momentum can also be shown in different forms, which will be used in further derivations. Multiplying (3.3-21) by Aw yields

$$m\left(\frac{\partial w}{\partial t} + w\frac{\partial w}{\partial z}\right) + Aw\Delta P_G = -Aw\frac{\partial P}{\partial z} - Aw\Delta P_F \quad . \qquad (3.3\text{-}22)$$

Substituting w = mv/A for w in Equation (3.3-21) and referring to the continuity equation yields the form of momentum equation that includes the gradient of mass flow rate m with respect to time

$$\frac{1}{A}\frac{\partial m}{\partial t} + \frac{1}{A^2}\frac{\partial(m^2 v)}{\partial z} = -\frac{\partial P}{\partial z} - \Delta P_G - \Delta P_F \quad . \qquad (3.3\text{-}23)$$

Finally, to analyze pressure change along the streamline of fluid the following form of the equation for momentum conservation is useful

$$\frac{1}{Av^2}\frac{\partial(mv^2)}{\partial t} + \frac{m^2}{A^2}\frac{\partial v}{\partial z} = -\frac{\partial P}{\partial z} - \Delta P_G - \Delta P_F \quad . \qquad (3.3\text{-}24)$$

Equations (3.3-21) - (3.3-24) are only different forms of the well-known Navier-Stokes equation for the one-dimensional flow of real fluid, in which the dynamic coefficient of viscosity μ differs from zero, i.e. in which friction is not neglected. The most similar in form to the Navier-Stokes equation is Equation (3.3-21), which becomes identical to it if it is divided by density ρ.

$$-\frac{1}{\rho}\frac{\partial P}{\partial z} = F_z + \frac{Dw}{dt} - Z \quad . \tag{3.3-25}$$

$$F_z = \frac{f}{\rho} \quad \ldots \text{ friction force per unit mass in the direction of the z axis}$$

$$\frac{Dw}{dt} \quad \ldots \text{ complete (material) gradient of velocity w change along the z axis}$$

$$Z = - g \frac{\partial h}{\partial z} \ldots \text{ gravitation force per unit mass in the direction of the z axis}$$

For ideal fluids (to which real gases and vapors with small coefficients of viscosity μ and high flow velocities come close) $\mu = 0$ and Equation (3.3-25) becomes the well-known Euler equation for ideal fluid

$$-\frac{1}{\rho}\frac{\partial P}{\partial z} + Z = \frac{Dw}{dt} \tag{3.3-26}$$

The extended Bernoulli's equation for the unsteady flow of viscous (real) liquid ($\mu \neq 0$, $\rho = $ const.) is also easy to derive from Equation (3.3-25). If the whole equation is multiplied by dz, after integration with respect to z from cross-section 1 to cross-section 2, we get

$$\frac{1}{\rho}\int_{P_1}^{P_2}\partial P + g\int_{h_1}^{h_2}\partial h + \int_{w_1}^{w_2}\partial(\frac{w^2}{2}) + \int_{z_1}^{z_2}\frac{\partial w}{\partial t}dz + \frac{\Delta P_F}{\rho}\int_{z_1}^{z_2}\partial z = 0 \quad .$$

$$\frac{1}{\rho}(P_2 - P_1) + g(h_2 - h_1) + (\frac{w_2^2}{2} - \frac{w_1^2}{2}) + \int_{z_1}^{z_2}\frac{\partial w}{\partial t}dz + \frac{\Delta P_F}{\rho}(z_2 - z_1) = 0$$

$$\tag{3.3-27}$$

In the flow of an **incompressible** liquid the gradient $\partial w/\partial t$ is not a function of z (A = const.) and equals dw/dt, so the integral in the last equation is

$$\frac{dw}{dt}(z_2 - z_1) = - \frac{dw}{dt}L \quad .$$

Rearrangement yields the well-known form of Bernoulli's equation for real liquid in unsteady conditions

$$\frac{P_1}{\rho} + gh_1 + \frac{w_1^2}{2} = \frac{P_2}{\rho} + gh_2 + \frac{w_2^2}{2} + \frac{dw}{dt}L + \frac{\Delta P_F}{\rho}L \ . \qquad (3.3\text{-}28)$$

Multiplying the above equation by ρ, and for $w = m/A\rho$, we get

$$(P_1 - P_2) - \underbrace{\Delta P_F L}_{\delta P_C} + \underbrace{\rho g(h_1 - h_2)}_{\delta P_G} + \frac{w_1^2 - w_2^2}{2}\rho = \frac{L}{A}\frac{dm}{dt} \ . \qquad (3.3\text{-}29)$$

In the horizontal pipe of constant cross-section the last two terms on the left-hand side of the last equation are lost and it becomes equal to Equation (2.2-1.4) from the chapter on lumped parameters.

EQUATION FOR THE CONSERVATION OF ENERGY

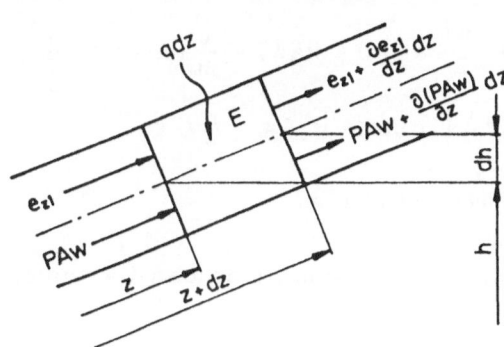

Fig. 3.3-4 Sketch for deriving energy equation

The law for the conservation of energy (thermal and mechanical) expresses the fact that the difference between the amounts of energy brought to and led from the fluid particle dM, in the elementary volume dV during time dt, equals the change of the total amount of energy dE contained in dV

$$e_i dt - e_o dt = dE \qquad (3.3\text{-}30)$$

A kilogram of flowing fluid mass possesses three forms of energy: internal thermal energy u, kinetic energy $w^2/2$ and mechanical potential energy gh. Therefore, the energy flow rate brought through cross-section z into dV is $e_{z1} = m(u+w^2/2+gh)$ J/s. The power necessary to inject the mass of fluid m, under pressure P, into section z is $e_{z2} = PAw$ J/s. As the fluid flows out of cross-section $z+dz$ it takes with it a total energy flow rate of

$e_z + de_z$ J/s, and to eject fluid out of the same cross-section the power $PAw + d(PAw)$ is needed. Finally, a heat flow rate qdz J/s is brought into dV from the outside through the pipe wall. What has just been said can be given the following mathematical form

$$e_i = e_z + Q = \underbrace{m(u + \frac{w^2}{2} + gh)}_{e_{z1}} + \underbrace{PAw}_{e_{z2}} + qdz \quad . \qquad (3.3\text{-}31)$$

$$e_o = e_z + de_z = e_z + \frac{\partial}{\partial z}(m(u + \frac{w^2}{2} + gh) + PAw)dz \quad . \qquad (3.3\text{-}32)$$

$$dE = dM(u + \frac{w^2}{2} + gh) = A\rho(u + \frac{w^2}{2} + gh)dz \quad . \qquad (3.3\text{-}33)$$

Substituting the last three equations into (3.3-30) and dividing by dz yields

$$q - \frac{\partial}{\partial z}(m(u + \frac{w^2}{2} + gh) + PAw) = \frac{\partial}{\partial t}(A\rho(u + \frac{w^2}{2} + gh)) \quad . \qquad (3.3\text{-}34)$$

Referring to (3.3-10) there is $Aw = mv$, so

$$mu + PAw = m(u + Pv) = mi \quad . \qquad (3.3\text{-}35)$$

The term ρu on the right-hand side can be written

$$\rho u = \rho i - P \quad . \qquad (3.3\text{-}36)$$

Substituting (3.3-35) and (3.3-36) into (3.3-34) and differentiating yields

$$q - \frac{\partial m}{\partial z}(i + \frac{w^2}{2} + gh) - m(\frac{\partial i}{\partial z} + w\frac{\partial w}{\partial z} + g\frac{\partial h}{\partial z}) =$$

$$= A\frac{\partial \rho}{\partial t}(i + \frac{w^2}{2} + gh) + A\rho(\frac{\partial i}{\partial t} + w\frac{\partial w}{\partial t}) - A\frac{\partial P}{\partial t} \qquad (3.3\text{-}37)$$

Referring to continuity equation (3.3-9), the terms $-i\partial m/\partial z$ and $iA\partial\rho/\partial t$ cancel out, so Equation (3.3-37), with reference to (3.3-10), can be rearranged into this final form

$$q = A\rho\frac{\partial i}{\partial t} - A\frac{\partial P}{\partial t} + m\frac{\partial i}{\partial z} + m(\frac{\partial w}{\partial t} + w\frac{\partial w}{\partial z}) + Aw\Delta P_G \quad . \qquad (3.3\text{-}38)$$

The last term of this equation was obtained, using notation from Equation (3.3-21), in the following manner

$$mg\frac{\partial h}{\partial z} = Aw\rho g\frac{\partial h}{\partial z} = Aw\Delta P_G \quad . \tag{3.3-39}$$

Equation (3.3-38) represents one of the possible forms of the law for energy conservation. As in the preceding examples, it is advantageous to show this law in different, more suitable, forms. One of them is obtained using the balance of momentum. It can be shown that the last two terms on the right-hand side of Equation (3.3-38) equal the left-hand side of Equation (3.3-22), so substituting the right-hand side of Equation (3.3-22) into (3.3-38) yields

$$(\frac{\partial i}{\partial t} + w\frac{\partial i}{\partial z}) - v(\frac{\partial P}{\partial t} + w\frac{\partial P}{\partial z}) = \frac{1}{A\rho}(q + Aw\Delta P_F) \quad , \tag{3.3-40}$$

$$\frac{Di}{dt} - v\frac{DP}{dt} = q_u \quad \frac{J}{kgs} \quad . \tag{3.3-41}$$

The total heat flow rate brought to one kilogram of fluid mass is denoted q_u, and it equals the sum of heat flow rate brought from the outside through the pipe wall $q/A\rho$ and the heat flow rate released by friction $w\Delta P_F/\rho$.

Equation (3.3-36) yields $i = u + vP$, so the energy conservation Equation (3.3-41) becomes

$$\frac{Du}{dt} + P\frac{Dv}{dt} = q_u \quad . \tag{3.3-42}$$

Multiplying (3.3-42) by dt yields confirmation of the first main law of thermodynamics, according to which the total heat introduced, $dQ = q_u dt$, is spent on an increase of the fluid's thermal energy and on mechanical work

$$dQ = Du + PDv \quad \frac{J}{kg} \quad . \tag{3.3-43}$$

According to Equation (3.3-14), the total change $Df(z, t)$ equals

$$Df(z, t) = (\frac{\partial f}{\partial t} + w\frac{\partial f}{\partial z})dt \quad . \tag{3.3-44}$$

It is clear that the law for energy conservation should satisfy the second main law of thermodynamics as well, whose generally-known form for the elementary fluid particle is

$$ds = \frac{1}{T}(di - vdP) \quad . \tag{3.3-45}$$

In this case of spatial distribution for fluid flow the law given by Equation (3.3-45) remains the same, the only difference being that a material, substantial increase D is used instead of change d. Therefore

$$\frac{Ds}{dt} = \frac{1}{T}(\frac{Di}{dt} - v\frac{DP}{dt}) \quad . \tag{3.3-46}$$

Now the expression on the left-hand side of Equation (3.3-40), or (3.3-41), is replaced by Ds/dt, so the law for energy conservation, expressed through entropy change, becomes

$$\frac{\partial s}{\partial t} + w\frac{\partial s}{\partial z} = \frac{1}{T}(\frac{1}{A\rho}q + \frac{w}{\rho}\Delta P_F) \quad . \tag{3.3-47}$$

The equations for mass, energy and momentum conservation that we derived are not sufficient to obtain a solution in the case of unsteady fluid flow. They contain four unknown variables that change in time and in space (mass flow rate m, specific volume v, pressure P and enthalpy and/or entropy s) so the three equations obtained are not enough for a solution. That system of equations must be completed by the thermodynamic equation of state that relates any state equation with two others. Which form that equation will take (whether $v = v(P, s)$, $\rho = \rho(P, i)$, i = i(P,s), i = i(P, v) or s = s(P, i), s = s(P, v) and so on) depends on which variables from the conservation equations are selected for direct numerical calculation (integration). Thus, for example, numerical integration or solving the system of equations is possible after expressing $\partial\rho/\partial t$, $\partial m/\partial t$ and . s/∂t from Equations (3.3-9), (3.3-23) and (3.3-47). After discretization with respect to the spatial variable, and knowing the initial state, the first step of integration yields new values for ρ, m and s. However, before the second step of integration can be undertaken, new values for v, T and P must also be known, because those variables are in the system of equations being integrated. While v is given directly by $v = 1/\rho$, P and t must be determined from state equations $P = P(\rho, s)$ and $T = T(\rho, s)$. Only after these variables have been found is it possible to begin the second step of integration.

In practice, since m, ϑ and P are standard control variables, we usually try to get a system from the equations for conservation that contains differentiation with respect to time of mass flow rate, temperature (i.e. enthalpy or entropy) and pressure. For the needs of further dynamic analysis, the following model contains derivatives with respect to time of flow rate, entropy and pressure. Thus it is necessary to obtain, in an analytic and differentiable form the state equation

$$v = v(P, s) \quad . \tag{3.3-48}$$

The upper equation easily yields $\partial P/\partial t$. From

$$\frac{\partial v}{\partial t} = \underbrace{\frac{\partial v}{\partial P} \frac{\partial P}{\partial t}}_{\gamma_A} + \underbrace{\frac{\partial v}{\partial s} \frac{\partial s}{\partial t}}_{\gamma_B} \tag{3.3-49}$$

whence

$$\frac{\partial P}{\partial t} = \frac{1}{\gamma_A} \left(\frac{\partial v}{\partial t} - \gamma_B \frac{\partial s}{\partial t} \right) \quad . \tag{3.3-50}$$

γ_A and γ_B are derived from Equation (3.3-48) and in the general case they also are analytic functions of pressure and entropy.

$$\gamma_A = \left(\frac{\partial v}{\partial P} \right)_S = \gamma_A(P, s) \quad . \tag{3.3-51}$$

$$\gamma_B = \left(\frac{\partial v}{\partial s} \right)_P = \gamma_B(P, s) \quad . \tag{3.3-52}$$

(The variable γ_A is related to the velocity of sound in fluid $\gamma_A = -v^2/c^2$.)

Finally the model sought for is obtained from Equations (3.3-23), (3.3-47) and (3.3-50) with the help of the continuity equation, and it comprises a system of three first-order hyperbolic PDE.

$$\frac{\partial m}{\partial t} = - A \frac{\partial P}{\partial z} - \frac{2mv}{A} \frac{\partial m}{\partial z} - \frac{m^2}{A} \frac{\partial v}{\partial z} - A(\Delta P_G + \Delta P_F) \quad , \tag{3.3-53}$$

$$\frac{\partial s}{\partial t} = v \left(\frac{1}{AT} q - \frac{m}{A} \frac{\partial s}{\partial z} + \frac{mv}{AT} \Delta P_F \right) \quad , \tag{3.3-54}$$

$$\frac{\partial P}{\partial t} = \frac{1}{\gamma_A} \left[\underbrace{\frac{v^2}{A} \frac{\partial m}{\partial z}}_{\frac{\partial v}{\partial t}} - \gamma_B \underbrace{v \left(\frac{1}{AT} q - \frac{m}{A} \frac{\partial s}{\partial z} + \frac{mv}{AT} \Delta P_F \right)}_{\frac{\partial s}{\partial t}} \right] . \tag{3.3-55}$$

These three equations must necessarily be accompanied by the thermodynamic state equations (3.3-48), (3.3-51) and (3.3-52). Unlike in the preceding section where the derived models were in most cases linear after certain assumptions had been made, the model shown by the above system is nonlinear. This nonlinearity is not reflected only by the fact that it contains the squares m^2 and v^2, but also by the products of variables, for example $mv\partial m/\partial z$ and so on.

The selection of variables m, s and P is only one possible selection. Here it was made because we wanted to analyze the eigenvalues of the process of disturbance propagation through the fluid. Thanks to the fact that the energy equation is expressed by entropy change, it is possible to separate the hydrodynamic (fast, and thus considered adiabatic) propagation of disturbances in pressure and flow rate from the slower transport propagation of disturbances in the thermal state of the fluid.

The following must also be said concerning the system of equations (3.3-53) - (3.3-55), which we have called a model. It describes momentum, energy and mass transfer (i.e. changes of mass flow rate, entropy and pressure) in an arbitrary fluid and does not include any special limitations (except the assumptions made at the beginning) concerning the specific properties of the processes themselves, the characteristics of the object in which they occur or the conditions under which they occur. Therefore, this system of equations can be used for any arbitrary, one-dimensional flow and heat process (transport of water and other liquids, gaspipes, evaporators and steam generator superheaters, pneumatic pipes and so on).

All technical processes occur in systems with specific geometrical characteristics, in various devices, apparatus and plants whose size limits the domain of change of the spatial coordinate in differential transfer equations. The physical characteristics of the operating medium are also factors of limitation in the given system of equations. Besides these limitations, to solve the problem correctly it is also necessary to know how the system interacts with the environment at its boundary (boundary conditions) and the states of the process variables at the moment we consider the beginning of calculation (initial conditions). **The geometric and physical characteristics of the system, together with the boundary and the initial conditions, are a set that separates the observed specific process from the whole class of processes to which we can apply the system of equations (3.3-53) - (3.3-55)** .

Therefore, the mathematical model of a process under observation comprises a system of differential equations together with the geometric and physical characteristics. This is important to stress because in this book a certain mathematical record is often called the model of a process, and the mentioned additional conditions are not specially indicated. In such cases it is taken for granted that all those conditions are defined and known.

———————————

The model presented in the preceding example does not bear any resemblance to the hyperbolic PDE (3.3-1), nor does it contain second derivatives with respect to time and space. It is, however, a connected system of first-order hyperbolic PDE and with the help of certain mathematical transformations a second-order PDE in m or P can be obtained, which will be a more expanded form of the already shown simplest type in Equation (3.3-1). Since the mathematical apparatus that has been developed makes it much simpler to analyze a linear model, in the following example we will linearize the model (3.3-53) - (3.3-55) to obtain a linear second-order hyperbolic PDE for mass flow rate m. This linear model will be used to analyze the dynamic properties of disturbance propagation along pipes for fluid transport. Analogously to the preceding section where this was done for processes with equalization, it will be shown that the models from the section on lumped parameters are just reduced forms of the models shown here. We may also repeat that the linear model obtained reproduces faithfully all the essential dynamic properties of the real nonlinear process on condition that deviations from the steady state observed are small.

———————————

Example 3 Analysis of dynamic properties. Steady state. Linear model

For the process of one-dimensional fluid flow through a pipe of length L whose unsteady state changes are shown by the nonlinear model (3.3-53) - (3.3-55), determine the steady state and then:

a) derive a linear model for deviation from the steady state observed,

b) using the linear model obtained, analyze dynamic properties for:

1 - in the initial steady state the fluid is at rest and is not heated ($\overline{m}$ = 0, $\overline{q}$ = 0),

2 - the fluid flows through the pipe adiabatically ($\overline{q}$ = 0),

3 - the fluid flows through the pipe and is heated by a heat flow rate of constant density q along the pipe.

The following boundary conditions are given:

1st BC ... mass flow rate is given at the input cross-section $m(z=0,t) = m_0(t)$

2nd BC ... pressure is given at the output cross-section $P(z=L,t) = P_L(t)$

3rd BC ... fluid with entropy s_0 is supplied at the input cross-section, $s(z=0,t) = s_0(t)$

The cross-section A of the pipe is constant. All the variables are uniform in a cross-section. Pressure drops due to friction and gravitation are neglected, $\Delta P_F = \Delta P_G = 0$.

STEADY STATE CALCULATION

The equation of continuity yields

$$\frac{\partial \overline{m}}{\partial z} = 0 \ .$$ (3.3-56)

After setting the terms $\partial \overline{m}/\partial t$, ΔP_G and ΔP_F equal to zero, Equation (3.3-53) yields

$$\frac{\overline{m}^2}{A} \frac{\partial \overline{v}}{\partial z} = A \frac{\partial \overline{P}}{\partial z} \ .$$ (3.3-57)

Equation (3.3-54) yields

$$\frac{\partial \overline{s}}{\partial z} = \frac{1}{wAT\rho} \overline{q} \ .$$ (3.3-58)

The last three equations, with a given heat rate $\overline{q}$ and boundary conditions that are constant in the steady state $\overline{m}_0$, $\overline{P}_L$, $\overline{s}_0$, yield

$$\overline{m}(z) = \overline{m}_0 = \text{const.} \tag{3.3-59}$$

$$\overline{P}(z) = \overline{P}_L + \frac{\overline{m}_0^2}{A^2}(\overline{v}_L - \overline{v}(z)) \quad . \tag{3.3-60}$$

The further derivation of analytical expressions is coupled with difficulties because thermodynamic state variables are interrelated and without a knowledge of their relations in analytical form solutions cannot be reached except in special cases. Thus, for example, to determine the distribution of s along z, i.e. to find $\overline{s}(z)$, demands the integration of (3.3-58) with respect to z. However, (3.3-58) has temperature T in the denominator, which is also a function of z, $T(z)$.

One of the special cases where analytical expressions for all the state variables in the steady state can be obtained is the case when **fluid flows through the pipe and evaporates**, which makes it possible to assume that temperature is a function of the pressure of evaporation only, $T = T(\overline{P})$. We will now show how to determine the steady state for **the case of the evaporator**.

Before deriving the other analytical expressions we must point out that the second term on the right-hand side of Equation (3.3-60) is as a rule, very, very small (less than 0.01% $\overline{P}_L$) and can be neglected. If this is so, then $\overline{P}(z) = \overline{P}_L$, or

$$\frac{\partial \overline{P}}{\partial z} = 0. \tag{3.3-61}$$

Beside constant pressure, temperature is also constant along the evaporator, $T(z) = \text{const.}$ Now Equation (3.3-58) can be integrated, which yields

$$\overline{s}(z) = \overline{s}_0 + \frac{1}{m_0 T}\overline{q}z \quad . \tag{3.3-62}$$

To determine $\overline{v}(z)$ and $\overline{w}(z)$, it is useful to have $\overline{i}(z)$. Referring to (3.3-61), Equation (3.3-43) yields

$$\overline{i}(z) = \overline{i}_0 + \frac{\overline{q}}{m}z, \quad \overline{i}_0 = i(\overline{s}_0) \quad . \tag{3.3-63}$$

In the evaporation process we have

$$\frac{\partial \bar{v}}{\partial z} = \frac{\partial \bar{i}}{\partial z} \frac{\bar{v}'''}{\bar{r}} \; , \qquad\qquad \begin{array}{l} \bar{v}''' = \bar{v}'' - \bar{v}' = v'''(\bar{P}) \\ \bar{r} = \bar{i}'' - \bar{i}' = r(\bar{P}) \end{array} \qquad (3.3\text{-}64)$$

The given equation yields

$$\bar{v}(z) = \bar{v}_0 + \frac{\bar{v}'''}{\bar{r}} \frac{\bar{q}}{m_0} z \; . \qquad\qquad (3.3\text{-}65)$$

Finally, referring to Equation (3.3-10), it follows that

$$\bar{w}(z) = \bar{w}_0 + \frac{\bar{v}'''}{\bar{r}} \frac{\bar{q}}{A} z \; . \qquad\qquad (3.3\text{-}66)$$

Therefore, in the evaporator pipe it is possible (with the assumption $\bar{P}(z) = \text{const.}$) to obtain analytical expressions for all process variables in the steady state. A characteristics of the evaporator is that those variables (except for flow rate, pressure and temperature which are constant) change linearly along z.

In all other cases, when it is not possible simply to determine the analytical expressions for variables in the steady state, this must be done with the help of a computer and using iteration.

Only then, after obtaining a specific steady operating state, can coefficients in the linearized equations be calculated.

a) MODEL LINEARIZATION

The process of obtaining a linear model from the initial system of equations (3.3-53) - (3.3-55) is based on the assumption that variables P, s, v, m, T and q remain in the neighborhood of steady state $\bar{P}$, $\bar{s}$, $\bar{v}$, $\bar{m}$, $\bar{T}$, and $\bar{q}$ about which they can be expanded into a Taylor series in which higher-order terms are neglected. (The procedure of linearization is shown in more detail in the Appendix.) Here we will show the linearization of Equation (3.3-53), and for the other equations only the final linear expressions will be given. With $\Delta P_G = \Delta P_F = 0$, Equation (3.3-53) becomes

$$\frac{\partial m}{\partial t} = -A \frac{\partial P}{\partial z} - \frac{m^2}{A} \frac{\partial v}{\partial z} - \frac{2mv}{A} \frac{\partial m}{\partial z} \qquad\qquad (3.3\text{-}67)$$

The gradient $\partial m/\partial t$ depends on the other variables as follows

$$\frac{\partial m}{\partial t} = g \left(\frac{\partial P}{\partial z}, m, \frac{\partial v}{\partial z}, v, \frac{\partial m}{\partial z} \right) \; . \qquad\qquad (3.3\text{-}68)$$

Differentiating (3.3-67) with respect to the functions on which m depends (that functional dependence is given by the last equation) and replacing d with Δ yields

$$\frac{\partial \Delta m}{\partial t} = -A\frac{\partial \Delta P}{\partial z} - \frac{\overline{m}^2}{A}\frac{\partial \Delta v}{\partial z} - (\frac{2\overline{m}}{A}\frac{\partial \overline{v}}{\partial z} + \frac{2\overline{v}}{A}\frac{\partial \overline{m}}{\partial z})\Delta m - \qquad (3.3\text{-}69)$$

$$- \frac{2\overline{m}}{A}\frac{\partial \overline{m}}{\partial z}\Delta v - \frac{2\overline{m}\overline{v}}{A}\frac{\partial \Delta m}{\partial z} \quad .$$

To obtain the linearized left-hand side we made use of the following property, for small deviations, of the operator of differentiation

$$d(\frac{\partial m}{\partial t}) = \frac{\partial (dm)}{\partial t} \quad . \qquad (3.3\text{-}70)$$

Referring to (3.3-56) yields the final, linear equation for the gradient of small deviation of flow rate Δm with respect to time t

$$\frac{\partial \Delta m}{\partial t} = -A\frac{\partial \Delta P}{\partial z} - \frac{\overline{m}^2}{A}\frac{\partial \Delta v}{\partial z} - \frac{2\overline{m}}{A}\frac{\partial \overline{v}}{\partial z}\Delta m - \frac{2\overline{m}\overline{v}}{A}\frac{\partial \Delta m}{\partial z} \qquad (3.3\text{-}71)$$

Linearization of the other two equations of the model and of the equation of continuity yields

$$\frac{\partial \Delta s}{\partial t} = \frac{\overline{v}}{AT}\Delta q - \frac{\overline{vm}}{A}\frac{\partial \Delta s}{\partial z} - \frac{\overline{v}}{A}\frac{\partial \overline{s}}{\partial z}\Delta m - \frac{\overline{v}}{AT^2}\overline{q}\,\overline{\tau}_p\Delta P - \frac{\overline{v}}{AT^2}\overline{q}\,\overline{\tau}_s\Delta s \qquad$$
$$(3.3\text{-}72)$$

$$\frac{\partial \Delta P}{\partial t} = \frac{1}{\gamma_A}(\frac{\partial \Delta v}{\partial t} - \overline{\gamma}_B\frac{\partial \Delta s}{\partial t}) \quad , \qquad (3.3\text{-}73)$$

$$\frac{\partial \Delta v}{\partial t} = \frac{\overline{v}^2}{A}\frac{\partial \Delta m}{\partial z} \quad . \qquad (3.3\text{-}74)$$

The above equations represent the linear model demanded. To obtain the last two terms in Equation (3.3-72) it is necessary to have the state equation $T = T(P,s)$, from which follows

$$dT = \underbrace{\frac{\partial T}{\partial P}}_{\tau_p} dP + \underbrace{\frac{\partial T}{\partial s}}_{\tau_s} ds \quad . \qquad (3.3\text{-}75)$$

We must, nevertheless, say that the influence of those last two terms in (3.3-72) is small because their denominators have the very large value T^2.

The model obtained is a linear system of three first-order, hyperbolic, constant coefficient PDE with respect to time. For further needs the following notation must be introduced for those coefficients

$$K_1 = A, \quad K_2 = \frac{\overline{m}^2}{A}, \quad K_3 = \frac{2\overline{m}}{A}\frac{\partial \overline{v}}{\partial z}, \quad K_4 = \frac{2\overline{m}\overline{v}}{A} = 2\overline{w},$$

$$K_5 = \frac{\overline{v}^2}{A}, \quad K_6 = \frac{\overline{v}}{AT}, \quad K_7 = \frac{\overline{m}\overline{v}}{A} = \overline{w}, \quad K_8 = \frac{\overline{v}}{A}\frac{\partial \overline{s}}{\partial z},$$

$$K_9 = \frac{\overline{v}}{AT^2}\overline{q}\overline{\tau}_p, \quad K_{10} = \frac{1}{\gamma_A}, \quad K_{11} = \frac{\gamma_B}{\gamma_A}, \quad K_{12} = K_5 K_{10}$$ (3.3-76)

$$K_{13} = \frac{\overline{v}}{AT^2}\overline{q}\overline{\tau}_s \quad .$$

In the general case of flow and heating of an arbitrary fluid, all the coefficients except K_1 and K_2 are functions of the spatial variable

$$K_i = f_i(z), \quad i = 3, 13 \quad .$$ (3.3-77)

The above expressions for the coefficients K_i show that they are functions of the operating state in which linearization was performed, that they change depending on it, and since the coefficients of any particular partial derivative determine the dynamic properties of the process itself, those dynamic properties will also depend on the operating state, i.e. on the conditions under which the process occurs.

b) ANALYSIS OF DYNAMIC PROPERTIES

As part of the analysis of dynamics we will show here the transfer functions for mass flow rate (the expressions $\Delta M(z, s)/\Delta M_0(s)$ and $\Delta M(z,s)/\Delta P_L(s)$) and examine their poles and zeros, which together with the gain coefficients determine the transient process. That end can be reached in various ways, but here we will transform the initial model into a second-order hyperbolic PDE for mass flow rate (or pressure) and retain the first-order hyperbolic PDE for the transfer of entropy disturbances.

Before we begin work on obtaining transfer functions we must say that model (3.3-71) - (3.3-73) represents the propagation of **small** disturbances Δm, Δs and ΔP along the streamlines of flow. Earlier, in systems with lumped parameters, it was shown that small disturbances of flow and pressure propagate through the fluid with the velocity of sound (the waves of compression and expansion have that velocity), while changes in thermal state (propagation of the disturbance Δs) occur convectively (carried by the velocity of fluid flow). These are, therefore, dynamically two different processes: a fast process (for Δm and ΔP) and the propagation of entropy disturbances. It is known that pressure and flow disturbances propagate in adiabatic conditions ($\Delta s = 0$). The velocity of propagation is such that there is no time for heat to be exchanged with the environment in any part of the fluid during the time it takes for the disturbance to pass. Thanks to this, the analysis can be divided into investigations into the dynamics of fast and of slow processes, and Equations (3.3-71) and (3.3-73), which contain $\Delta s = 0$, can be analyzed separately from Equation (3.3-72). This is what we have done, but since this section is concerned with periodic processes, the analysis of the first-order hyperbolic PDE for changes of Δs will in most cases be left out.

If we differentiate Equation (3.3-71) with respect to t and replace the expression $\partial^2 \Delta v / \partial t \partial z$ in it with the equivalent value obtained from (3.3-74) after differentiation with respect to z, a second-order PDE is obtained for $\Delta m(z, t)$ which contains the partial derivative $\partial^2 \Delta P / \partial t \partial z$. Substituting (3.3-74) into (3.3-73) and differentiating (3.3-73) with respect to z yields that derivative, whose substitution yields the final expression describing changes in deviations $\Delta m(z, t)$

$$\frac{\partial^2 \Delta m}{\partial t^2} + K_3 \frac{\partial \Delta m}{\partial t} + (K_2 K_5 + K_1 K_5 K_{10}) \frac{\partial^2 \Delta m}{\partial z^2} +$$
$$+ (K_2 \frac{\partial K_5}{\partial z} + K_1 \frac{\partial K_{12}}{\partial z}) \frac{\partial \Delta m}{\partial z} + K_4 \frac{\partial^2 \Delta m}{\partial t \partial z} = 0 \quad . \tag{3.3-78}$$

It will be easier to understand this PDE if the physical meaning of the coefficients is given beside the partial derivatives. In the general case, all those coefficients are functions of the spatial coordinate and the equalities (3.3-76) yield

$$K_3 = K_3(z), \quad K_2 K_5 = \overline{w}^2(z), \quad K_4 = 2\overline{w}(z) \tag{3.3-79}$$

$$K_1 K_5 K_{10} = A \frac{\overline{v}^2}{A} \frac{1}{\gamma_A} = \overline{v}^2 (\frac{\partial P}{\partial v})_s = -\overline{c}^2 \quad .$$

$\overline{w}$ is the fluid flow velocity and $\overline{c}$ the velocity of sound in steady state. If we introduce the notation

$$f_1(z) = K_2 K_5 + K_1 K_5 K_{10} = \overline{w}^2(z) - \overline{c}^2(z) \qquad \left[\frac{m^2}{s^2}\right]$$

$$f_2(z) = K_2 \frac{\partial K_5}{\partial z} + K_1 \frac{\partial K_{12}}{\partial z} \qquad \left[\frac{m}{s^2}\right] \tag{3.3-80}$$

$$f_3(z) = K_4 = 2\overline{w}(z) \quad \left[\frac{m}{s}\right] \quad .$$

Equation (3.3-78) becomes

$$\frac{\partial^2 \Delta m}{\partial t^2} + K_3(z)\frac{\partial \Delta m}{\partial t} + f_1(z)\frac{\partial^2 \Delta m}{\partial z^2} + f_2(z)\frac{\partial \Delta m}{\partial z} + f_3(z)\frac{\partial^2 \Delta m}{\partial t \partial z} = 0 \qquad (3.3\text{-}81)$$

We have obtained a second-order hyperbolic PDE that resembles the basic and simplest Equation (3.3-1) from the beginning of this section and thus points to periodic changes of mass flow rate. Here we have left out the derivation of the general PDE for pressure change. It is not difficult to prove that the dynamic coefficients, equations, transfer functions, their poles and zeros and so on for pressure changes are equal to those for flow changes, which will be shown for the simplest cases.

b1) Fluid at rest without heating

This is a special case and the simplest, when the medium is at rest in a pipe of length L ($\overline{m} = 0$, $\overline{w} = 0$) and there is no heating along the pipe ($\overline{q} = 0$), so that (3.3-76) yields

$$K_2 = K_3 = K_4 = K_7 = K_8 = K_{13} = 0 \quad ,$$

$$\frac{\partial K_5}{\partial z} = \frac{\partial K_{12}}{\partial z} = 0.$$

$$\hspace{11cm} (3.3\text{-}82)$$

Thus Equation (3.3-81) obtains the form of the classical wave equation

$$\frac{\partial^2 \Delta m}{\partial t^2} - \overline{c}^2 \frac{\partial^2 \Delta m}{\partial z^2} = 0 \quad .$$

$$\hspace{11cm} (3.3\text{-}83)$$

Further mathematical procedure is identical to methods that were used in the preceding section for processes with equalization, and is reduced to the fact that after Laplace transformation of the initial PDE we get an ODE for $\Delta M(z, s)$ with respect to z in the s-domain, whose solution is a sum of exponential functions beside which are coefficients that are functions of the boundary conditions. On the boundaries $z = 0$ and $z = L$ these conditions are

1st BC, $z = 0$, $\Delta m(0, t) = \Delta m_0(t)$.

$$\hspace{11cm} (3.3\text{-}84)$$

2nd BC, $z = L$, $\Delta P(L, t) = \Delta P_L(t)$.

$$\hspace{11cm} (3.3\text{-}85)$$

From (3.3-73) and (3.3-74), in the case of adiabatic processes $\Delta s = 0$, the second boundary conditions becomes

$$\frac{\partial \Delta m(L, t)}{\partial z} = \frac{1}{K_5 K_{10}} \frac{\partial \Delta P(L, t)}{\partial t} \quad . \tag{3.3-86}$$

The solution of Equation (3.3-83) in the s-domain is

$$\Delta M(z, s) = C_1 e^{\frac{s}{c}z} + C_2 e^{-\frac{s}{c}z} \quad . \tag{3.3-87}$$

For conditions on the boundaries $z = 0$ and $z = L$, C_1 and C_2 are determined from Equation (3.3-87) and its derivative with respect to z (to satisfy the second BC shown in (3.3-86))

$$C_1 = \frac{\Delta M_0(s) e^{-\frac{L}{c}s} + \frac{\bar{c}}{K_5 K_{10}} \Delta P_L(s)}{e^{\frac{L}{c}s} + e^{-\frac{L}{c}s}} \quad , \tag{3.3-88}$$

$$C_2 = \frac{\Delta M_0(s) e^{\frac{L}{c}s} - \frac{\bar{c}}{K_5 K_{10}} \Delta P_L(s)}{e^{\frac{L}{c}s} + e^{-\frac{L}{c}s}} \quad . \tag{3.3-89}$$

The final expression for the change of mass flow rate at the output cross-section is obtained from (3.3-87) - (3.3-89)

$$\Delta M(L, s) = \underbrace{\frac{1}{ch\frac{L}{c}s}}_{G_1(s)} \Delta M_0(s) - \underbrace{\frac{A}{c} \frac{sh\frac{L}{c}s}{ch\frac{L}{c}s}}_{G_2(s)} \Delta P_L(s) \quad . \tag{3.3-90}$$

The poles of the transfer functions (whose physical meaning is that they represent natural frequencies of oscillation of the column of fluid) characterize the dynamic properties and it is not difficult to determine them by setting the denominators of $G_1(s)$ and $G_2(s)$ equal to zero

$$ch(\frac{L}{c}\sigma + \frac{L}{c}\omega j) = ch\frac{L}{c}\sigma \cos\frac{L}{c}\omega + j\,sh\frac{L}{c}\sigma \sin\frac{L}{c}\omega = 0.$$

Both the real and the imaginary parts will equal zero at the same time only if

$$\cos\frac{L}{c}\omega = 0 \quad \text{and} \quad \text{sh}\frac{L}{c}\sigma = 0 \ ,$$

whence the poles of $G_1(s)$ and $G_2(s)$ are

$$\sigma = 0 \ , \tag{3.3-91}$$

$$\omega_k = (2k - 1)\frac{\pi}{2}\frac{\bar{c}}{L},$$

$$= \pm(2k + 1)\frac{\pi}{2}\frac{\bar{c}}{L}, \quad k = 0, 1, 2, 3, \dots \tag{3.3-92}$$

Setting the numerator of $G_2(s)$ equal to zero yields the zeros of that transfer function

$$\sigma_0 = 0 \ , \tag{3.3-93}$$

$$\omega_{0k} = k\pi\frac{\bar{c}}{L}, \quad k = 0, \pm 1, \pm 2, \dots$$

$$= \pm k\pi\frac{\bar{c}}{L}, \quad k = 0, 1, 2, 3, \dots \tag{3.3-94}$$

The fact that the poles of the transfer functions are complex conjugate pairs whose real parts equal zero allows us to conclude that the flow rate $\Delta M_L(s)$, in the case of any disturbance from the boundary, will oscillate with frequencies ω_k permanently and without damping ($\sigma = 0$). This is the idealized case of a fluid at rest without energy exchange and neglecting all possible losses. The form of PDE that describes it is equal to the PDE for longitudinal vibrations of a rod or beam (3.3-4). This similarity is not unexpected because in the observed case, when there is no flow, the pipe in fact contains a "rod" of fluid with its elasticity modulus and velocity of small pressure disturbances propagation along the pipe.

Besides, it can be seen that in the case of a finite-dimensional space (limited space) the equations of conservation do not have a solution that would satisfy the given boundary conditions for any arbitrary frequency. Such solutions exist only for a completely determined spectrum of values given by Equations (3.3-91) and (3.3-92). In other words, in a space with a finite dimensions, oscillations are possible only for a specific set of frequencies (of which there are, true, infinitely many) which, since they uniquely describe the properties of that space, are called the natural frequencies.

These results show the main difference between the properties of these processes and properties of processes with equalization, where the poles can be real and negative, while here they are complex conjugate pairs.

It was mentioned earlier that the dynamics of pressure change along a pipe of fluid is described by the same PDE that describes the dynamics of flow rate. Now this will be shown. We will start from Equations (3.3-71) and (3.3-73), which, referring to (3.3-74) and (3.3-82), in the adiabatic case ($\Delta s = 0$) obtain the following simpler forms

$$\frac{\partial \Delta m}{\partial t} = - A \frac{\partial \Delta P}{\partial z} \quad , \tag{3.3-95}$$

$$\frac{\partial \Delta P}{\partial t} = \frac{\bar{v}^2}{A} \frac{1}{\gamma_A} \frac{\partial \Delta m}{\partial z} \quad . \tag{3.3-96}$$

Differentiating (3.3-95) with respect to z and (3.3-96) with respect to t yields

$$\frac{\partial^2 \Delta P}{\partial t^2} = \frac{\bar{v}^2}{A} \frac{1}{\gamma_A} \frac{\partial^2 \Delta m}{\partial t \partial z} = - A \frac{\bar{v}^2}{A} \frac{1}{\gamma_A} \frac{\partial^2 \Delta P}{\partial z^2} \quad ,$$

$$\frac{\partial^2 \Delta P}{\partial t^2} = \bar{c}^2 \frac{\partial^2 \Delta P}{\partial z^2} \tag{3.3-97}$$

This equation is completely the same as (3.3-83), which shows that the dynamic properties of processes of flow rate and pressure disturbances propagation are identical.

For the given boundary conditions, using the same procedure as before, we obtain

$$\Delta P_0(s) = \underbrace{\frac{1}{ch\frac{L}{c}s}}_{G_3(s)} \Delta P_L(s) + \underbrace{\frac{\bar{c}}{A} \frac{sh\frac{L}{c}s}{ch\frac{L}{c}s}}_{G_4(s)} \Delta M_0(s) \tag{3.3-98}$$

Comparing G_3 and G_4 with G_1 and G_2 from (3.3-90) it is obvious that the poles and zeros derived there will satisfy here also for dynamic pressure changes at the input cross-section $\Delta P_0(s)$.

Equations (3.3-90) and (3.3-98) can also be shown in matrix form which will later be used for a comparison with the models from Section (2.2-3) on processes with lumped parameters.

$$
\begin{bmatrix} P_0(s) \\ \\ \Delta M_L(s) \end{bmatrix} = \begin{bmatrix} \dfrac{1}{ch\frac{L}{c}s} & \dfrac{\bar{c}}{A}\dfrac{sh\frac{L}{c}s}{ch\frac{L}{\bar{c}}s} \\ \\ -\dfrac{A}{\bar{c}}\dfrac{sh\frac{L}{c}s}{ch\frac{L}{c}s} & \dfrac{1}{ch\frac{L}{c}s} \end{bmatrix} \cdot \begin{bmatrix} \Delta P_L(s) \\ \\ \Delta M_0(s) \end{bmatrix} . \qquad (3.3\text{-}99)
$$

b2) Fluid flow without heating

The case of adiabatic fluid flow through pipes is very frequent in technical practice and is met in the transport of liquids, gases and vapors. Unlike in the preceding case, now the velocity and mass flow rate are not equal to zero in the steady state observed ($\bar{m} \neq 0$, $\bar{w} \neq 0$), but $\bar{q} = 0$ still holds. According to (3.3-76), here we will have

$$
K_3 = K_8 = K_9 = K_{13} = 0, \quad \frac{\partial K_5}{\partial z} = \frac{\partial K_{12}}{\partial z} = 0 \quad, \qquad (3.3\text{-}100)
$$

and the initial and general PDE (3.3-81) will obtain this special form

$$
\frac{\partial^2 \Delta m}{\partial t^2} + (\bar{w}^2 - \bar{c}^2)\frac{\partial^2 \Delta m}{\partial z^2} + 2\bar{w}\frac{\partial^2 \Delta m}{\partial t\partial z} = 0 \quad. \qquad (3.3\text{-}101)
$$

For the same boundary conditions and with the same procedure as in the preceding case changes of mass flow rate at the output cross-section are

$$
\Delta M_L(s) = \underbrace{\frac{2\bar{c}e^{-\frac{2L\bar{w}}{\bar{w}^2-\bar{c}^2}s}}{(\bar{w}+\bar{c})e^{-\frac{L}{\bar{w}-\bar{c}}s} - (\bar{w}-\bar{c})e^{-\frac{L}{\bar{w}+\bar{c}}s}}}_{G_1(s)} \Delta M_0(s) -
$$

$$
\qquad\qquad (3.3\text{-}102)
$$

$$
- A\underbrace{\frac{\frac{\bar{c}^2-\bar{w}^2}{\bar{c}^2}(e^{-\frac{L}{\bar{w}-\bar{c}}s} - e^{-\frac{L}{\bar{w}+\bar{c}}s})}{(\bar{w}+\bar{c})e^{-\frac{L}{\bar{w}-\bar{c}}s} - (\bar{w}-\bar{c})e^{-\frac{L}{\bar{w}+\bar{c}}s}}}_{G_2(s)}\Delta P_L(s) \quad.
$$

Setting the denominator (which is common to both the transfer functions) equal to zero in the case of **subcritical** flow ($\overline{w} < \overline{c}$) gives the poles of $G_1(s)$ and $G_2(s)$

$$s = \frac{\overline{c}^2 - \overline{w}^2}{2\overline{c}L} \ln\left|\frac{\overline{w} - \overline{c}}{\overline{w} + \overline{c}}\right| \pm \frac{\overline{c}^2 - \overline{w}^2}{2\overline{c}L}(2k + 1)\pi j, \quad k = 0, 1, 2, 3, \ldots$$
$$(3.3\text{-}103)$$

The zeros of the transfer functions $G_1(s)$ and $G_2(s)$ are obtained by setting their numerators equal to zero:

zeros of $G_1(s)$ ... in the s-plane they lie on the "straight line" $\sigma_{01} = -\infty$

$$(3.3\text{-}104)$$

zeros of $G_2(s)$ ... $\sigma_{02} = 0$, $\omega_{02k} = \pm k\pi \dfrac{\overline{c}^2 - \overline{w}^2}{\overline{c}L}$ \qquad (3.3\text{-}105)

Unlike in the preceding case b1), where the real part of the set of eigenvalues (the poles of the transfer functions) equalled zero, i.e. when the transient process occurred without damping, in this case σ is different from zero and after any disturbance at the pipe ends, the periodic changes $\Delta m_L(t)$ are damped. Their oscillation amplitudes (although idealization was performed by taking $\Delta P_F = \Delta P_G = 0$) converge to zero with time. Also characteristic of the natural frequencies is that both the lowest and the highest are equally damped (the same σ).

The appearance of the real part that is not equal to zero is a very important characteristic of adiabatic flow. The transfer functions obtained differ essentially from those in the case when the flow process was observed as a system with lumped parameters. If $R = 0$ is substituted into the transfer functions shown by Equation (2.2-3.40) (which corresponds to the loss, neglected here also, resulting from friction, $\Delta P_F = 0$), the classical case of undamped oscillation is obtained, with characteristic complex conjugate pairs of poles on the imaginary axis. Obviously, the differences are of an essential nature. For practical purposes, however, it can be shown that if certain assumptions are made the model shown by Equation (3.3-102) can be translated into forms that have their equivalent in processes with lumped parameters. In the usual processes of fluid flow (which is especially true of liquids) the difference between the velocity of flow w and the speed of sound propagation c are very great, so that w in Equation (3.3-102) and (3.3-103) can be neglected. With this assumption just derived PDE, transfer functions, poles and zeros are reduced to the results in case b1), so that for w = 0 the formulas (3.3-91) - (3.3-94) will satisfy here also, and so will the matrix transfer function (3.3-99). If the expressions for hyperbolic functions from (3.3-99) are expanded into a series and all the terms of third and higher order neglected, introducing the coefficients

$$\frac{L}{c} = T_L, \quad \frac{\bar{c}}{A} = Z, \quad IC = T_L^2, \quad \frac{T_L}{Z} = C,$$

yields the following first, roughest approximation of the matrix transfer functions

$$\begin{bmatrix} P_0(s) \\ \\ \Delta M_L(s) \end{bmatrix} = \begin{bmatrix} \dfrac{1}{\dfrac{IC}{2}s^2+1} & \dfrac{Is}{\dfrac{IC}{2}s^2+1} \\ \\ \dfrac{-Cs}{\dfrac{IC}{2}s^2+1} & \dfrac{1}{\dfrac{IC}{2}s^2+1} \end{bmatrix} \begin{bmatrix} \Delta P_L(s) \\ \\ \Delta M_0(s) \end{bmatrix}. \qquad (3.3\text{-}106)$$

A form has been obtained that corresponds completely to the form of matrix transfer functions shown by Equation (2.2-3.40). The difference is that when flow is observed as a process with lumped parameters (Section 2.2-3), the constant beside s^2 is twice as large as the constant obtained here with the first approximation of the hyperbolic functions.

However, it must be born in mind that the similarity between Equations (2.2-3.40) and (3.3-106) has been obtained after neglecting flow velocity w. From this aspect, the model shown by (2.2-3.40) is limited to processes with small flow velocities. If the flow velocities w are large and of a similar order of magnitude as the speed of sound c, the initial equation (in the case of lumped parameters also) must be Equation (3.3-102) from which simpler forms, or finite-dimensional models, are obtained by expanding the exponential terms into a series and neglecting the higher-order terms in that series.

We can use a procedure analogous to the one used to obtain Equation (3.3-102) to get a relation between pressure changes in the cross-section z = 0 and changes on the boundaries

$$\Delta P_0(s) = \underbrace{\frac{2\bar{c}}{(\bar{w}+\bar{c})e^{-\frac{L}{w-c}s} - (\bar{w}-\bar{c})e^{-\frac{L}{w+c}s}}}_{G_3(s)} \Delta P_L(s) +$$

$$+ \frac{\bar{c}^2}{A} \underbrace{\frac{e^{-\frac{L}{w-c}s} - e^{-\frac{L}{w+c}s}}{(\bar{w}+\bar{c})e^{-\frac{L}{w-c}s} - (\bar{w}-\bar{c})e^{-\frac{L}{w+c}s}}}_{G_4(s)} \Delta M_0(s) \quad .$$

(3.3-107)

The poles and zeros of the transfer function $G_3(s)$ and $G_4(s)$ are equal to the poles and zeros determined with the help of (3.3-103) - (3.3-105). Like before, here also Equation (3.3-107) will be transformed in the boundary case, when w converges to zero, into Equation (3.3-98) obtained in the case when the fluid was at rest, b1).

b3) Fluid flow and heating

Studying the dynamic properties in this most complicated case, when the fluid flows and is heated at the same time by a heat flow of density $\bar{q}$ along the pipe (i.e. when there is $\bar{m} \neq 0$, $\bar{q} \neq 0$), is made much more difficult by the fact that the initial PDE (3.3-81) is now an equation with space-variable coefficients: $K_3(z)$, $f_1(z)$, $f_2(z)$ and $f_3(z)$. In practice it is no longer possible to obtain a solution in a closed form, i.e. in the form of analytical expressions, which was easy in the preceding cases b1) and b2). The coefficients are so complicated that it is necessary to use a digital computer and try to reach a satisfactory solution using the relevant approximations. Therefore, we will not enter further into the analysis of this general case here. The following numerical example, however, will show the results obtained for a special case, the analysis of the dynamic properties of a steam generator evaporator.

Nevertheless, before we begin to present the following example, this remark must be made to make the analysis complete. In the last example we examined only the dynamics of fast processes of pressure and flow rate disturbance propagation and completely left out an analysis of the dynamics of slow processes of entropy disturbance propagation, which was given by the first-order hyperbolic PDE (3.3-72). This type of equation was analyzed in Section 3.1, and since our purpose here was to examine periodic processes, we did not pay any attention to slow processes. It is necessary however, to point out that every disturbance includes simultaneous changes of mass flow rate and entropy and pressure, so this separation of fast periodic processes from slower convective processes of entropy disturbance propagation makes sense only within the framework of the theoretical analysis carried out here.

Example 4 Calculations of dynamic characteristics of an evaporator

The evaporator pipe of a steam generator of length L = 28 m and diameter d = 0.03 m is heated by a heat flow rate of q = 4688 J/ms density per unit length. Through the pipe flows m = 0.4783 kg/s of evaporating fluid (water-steam) that is saturated and at operating pressure 135 bar at the input cross-section. At the output cross-section, for a given q, it is a mixture with x = 0.25 steam quality. Determine the eigenvalues (the zeros of the transfer function denominator) that characterize the dynamics of change in mass flow rate.

After introducing some approximations, the solution demanded will be obtained numerically. This is because Equation (3.3-81) now has variable coefficients and cannot be solved analytically. In the specific case of the evaporator, K_3 depends only on the steam quality x, and for a given x it is constant along the whole evaporator. The equalities (3.3-79) and (3.3-80) show that coefficients f_1, f_2 and f_3 are functions of coordinate z, but also of steam quality x. Figure 3.3-5 shows how those coefficients change along the evaporator pipe depending on the degree of evaporation. (Here we must mention that the numerical program used to calculate these coefficients contained all the demanded equations for calculating the steady state (3.3-59) - (3.3-66), and also the thermodynamic state functions, for example (3.3-48), (3.3-51), (3.3-52).)

It is obvious that the dependence of these coefficients on z increases with the increase of the degree of evaporation, and that when x converges to zero they obtain the same values as the coefficients of the PDE (3.3-101) from the preceding example, i.e. $K_3 = 0$, $f_1 = \overline{w}^2 - \overline{c}^2$ = const., $f_2 = 0$ and $f_3 = 2\overline{w}$ = const. Figure 3.3-5 shows that changes in the coefficients f_i along z, depending on x, are not simple and for a suitable description, a second-order function should be used. But with that approximation an analytical solution in a closed form would not be possible. From this follows the natural conclusion that the solution, after making certain approximations, should be reached numerically. Therefore, to obtain transfer functions relating output flow rate with output pressure and input flow rate for the given $x = 0.25$, instead of variable coefficients along z, we will select for coefficients f_1, f_2 and f_3, the constants A, C, B and use them to solve the task demanded, i.e. to determine the eigenvalues. These constants are also the mean values shown in Figure 3.3-5 by broken lines.

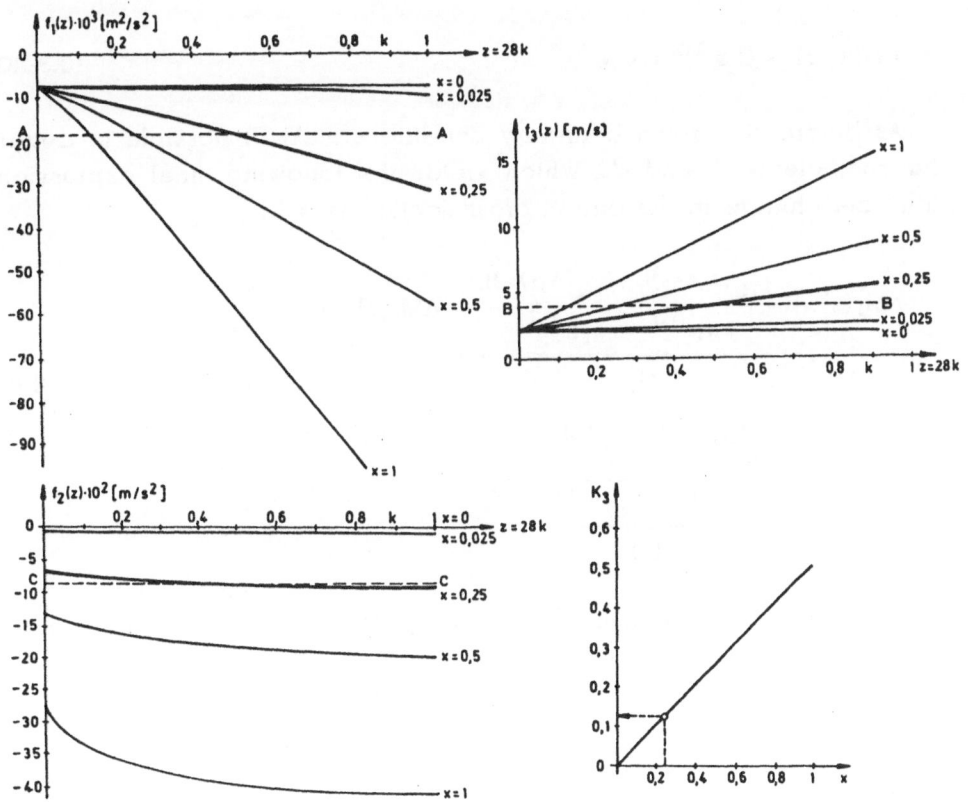

Fig. 3.3-5 Changes of the coefficients of PDE (3.3-81) along the evaporator depending on the output steam quality x

The introduction of those constants turns Equation (3.3-81) into a PDE which, after Laplace transformation with respect to time, becomes the following ODE with respect to z in the s-domain

$$A\frac{d^2\Delta M(z, s)}{dz^2} + (C + Bs)\frac{d\Delta M(z, s)}{dz} + (s^2 + K_3 s)\Delta M(z, s) = 0 \quad . \quad (3.3\text{-}108)$$

This equation is solved in the usual manner, by substituting $\Delta M(z, s) = \overline{C}e^{\lambda s}$. This leads to a characteristic equation whose roots are no longer simple expressions, but

$$\lambda_{1,2} = \frac{-(Bs+C) \pm \sqrt{(B^2-4A)s^2 + (2BC-4AK_3)s + C^2}}{2A} \quad . \quad (3.3\text{-}109)$$

With these $\lambda_{1,2}$ there is

$$\Delta M(z, s) = \overline{C}_1 e^{\lambda_1 z} + \overline{C}_2 e^{\lambda_2 z} \quad . \quad (3.3\text{-}110)$$

As before, the given boundary conditions make it possible to determine the coefficients $\overline{C}_1$ and $\overline{C}_2$, which yields the following, final expression for flow rate change in the output cross-section ($z = L$)

$$\Delta M_L(s) = \underbrace{\frac{\lambda_1 e^{(\lambda_1+\lambda_2)L} - \lambda_2 e^{(\lambda_1+\lambda_2)L}}{\lambda_1 e^{\lambda_1 L} - \lambda_2 e^{\lambda_2 L}}}_{G_1(s)} \Delta M_0(s) \; +$$

$$+ \; \underbrace{\frac{\frac{s}{K_{12}}(e^{\lambda_1 L} - e^{\lambda_2 L})}{\lambda_1 e^{\lambda_1 L} - \lambda_2 e^{\lambda_2 L}}}_{G_2(s)} \Delta P_L(s) \quad . \qquad (3.3\text{-}111)$$

The poles of the transfer functions are the eigenvalues demanded and they will be reached numerically, as the zeros of the transfer function denominator

$$N(s) = \lambda_1 e^{\lambda_1 L} - \lambda_2 e^{\lambda_2 L} = 0 \quad . \qquad (3.3\text{-}112)$$

A special program is needed for that purpose that will iteratively, applying Newton's method for the function of a complex variable $N(s)$, determine an arbitrary number of eigenvalues. Figure 3.3-6 shows those values both for the case $x = 0.25$ and for situations when unheated saturated water and unheated dry saturated steam flow through the same pipe (i.e. fluids with $x = 0$ and $x = 1$, and those pairs of eigenvalues are really the graphical representation of Equation (3.3-103)). Also, as in the

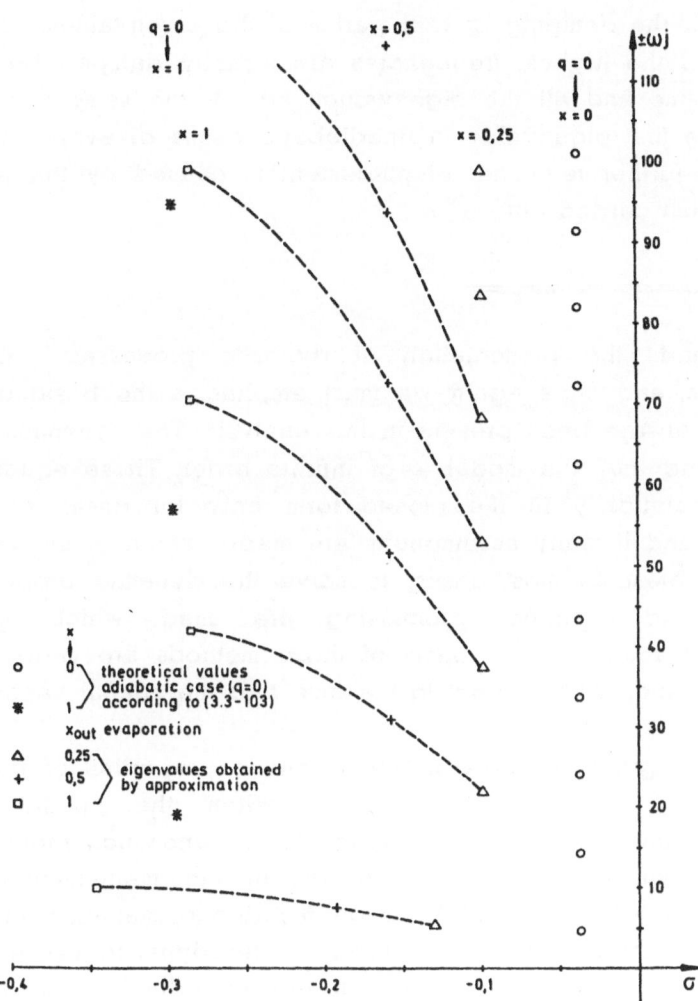

Fig. 3.3-6 Presentation of eigenvalues for cases of evaporation
(x_{out} = 0.25, 0.5, 1) and adiabatic flow with x = 0 and x = 1

case when there was x = 0.25 in the output cross-section, pairs of
eigenvalues obtained by approximation for the evaporator pipe with x =
0.5 and x = 1 at the output cross-section are shown. Several conclusions
can be drawn from Figure 3.3-6. First, the eigenvalues are complex
conjugate pairs, which again shows that changes of mass flow rate are
periodic. (It must be remarked that the ordinate shows both the positive
and the negative imaginary axis, which results in the complex-conjugate
pair being given by one point in that s-plane.) Second, the higher the

degree of evaporation the more leftwards the whole spectrum moves, i.e. the greater the damping or real part o of the eigenvalues. Third, both the lowest and the highest frequencies are equally damped because their o are the same and all the eigenvalues are on the same straight line. (All except the first eigenvalue in unadiabatic cases of evaporation, which is displaced further left. That displacement is caused by the approximation that was just carried out.)

This ends the presentation of dynamic processes with distributed parameters, and once again we must emphasize the basic characteristics that have always been present in this analysis. The dynamics is described by PDE and thus the model is of infinite order. Those equations can be solved analytically in the closed form only for cases of very simple geometry and if many assumptions are made, which mean an idealization of the problem. In most cases, to solve the dynamic problem numerical methods and computer processing are used, which usually gives satisfactory solutions. The basis of those methods are various procedures of discretization with respect to the time and the spatial variable.

There is great similarity in the dynamic properties of processes with lumped and distributed parameters. When the spatial variable is discretized the property of distribution is lost and such processes become processes with lumped characteristics. In the mathematical sense this means the order of the model has been reduced, making it finite, and now the model describes the basic (and in the dynamic sense the slowest) dynamic properties. The lower the degree of reduction, i.e. the higher the order of the model thus obtained, the better included the higher eigenvalues are (smaller time constants or greater natural frequencies). The responses shown by a model obtained by reduction deviate from the responses of the real distributed process in the first parts of the transient process (i.e. for smaller t), which is the result of the fact that the faster modes, in the dynamic sense, of the total response were not included in the model. The lower the degree of reduction, the smaller that " dynamic mistake" is.

DIFFERENTIAL EQUATIONS
AND STATE SPACE

One of the fundamental and most noticeable characteristics of all the processes that were analyzed on the pages of this book was that the mathematical tools used to describe their unsteady states were differential equations. These equations were described by different adjectives depending on their specific properties - ordinary, partial, nonlinear or linear. Their coefficients were constant or variable. We must remember that at the end we tried (and in most cases succeeded) to reduce all these different notations to one or several ordinary linear differential equations (OLDE). This equation or system of OLDE we tried to translate into the form of matrix state-space differential equations.

We will not discuss the accuracy, advantages or disadvantages of these mathematical tools. Many pages could be written about their successful use and the wide field in which these linear models can be applied. Here it is, perhaps, enough to say only that linear models are used in very many technical disciplines and that they have contributed to the practical and theoretical solution of many problems in these fields. Modern control theory is certainly one of the leading disciplines in the extensive use of the mentioned models, but useful and important results have also been obtained from their use in the theories of vibrations, acoustics, thermal and flow systems and in designing electrical circuits.

In the following lines, without the usual mathematical strictness, we will show some basic properties of differential equations, connections between one n-th-order DE and a system of first-order ODE, and how to present such a system in the form of (many times already mentioned) matrix state-space equations. Everyday technical practice has shown it to be useful to translate every existing model of dynamic processes into a system of first-order DE. (One of the more important reasons why this is so is that mathematical libraries of modern computers contain system programs specially developed for solving systems of first-order DE.)

Let it be possible to write the nonlinear (or linear) n-th-order DE as follows

$$x^{(n)} = f(t, x, x', x'', ..., x^{(n-1)}) \tag{A.1}$$

(In the above equation we did not specially emphasize that the function x and its derivatives are functions of the independent variable t, i.e. we did not write $x(t)$, $x'(t)$ ... and so on.)

The following substitution, i.e. the introduction of new functions

$$x = x_1,$$
$$x' = x_1' = x_2,$$
$$x'' = x_2' = x_3,$$
$$. . . . .$$
$$x^{(n-1)} = x'_{n-1} = x_n$$
$$x^{(n)} = x'_n = f(t, x_1, x_2, ..., x_n) \quad .$$

yields this system of n first-order DE

$$x_1' = x_2,$$
$$x_2' = x_3, \tag{A.2}$$
$$. . . . .$$
$$x'_n = f(t, x_1, x_2, ..., x_n) \quad .$$

The above system is only a special case or a form of the following system

$$x_1' = f_1 (t, x_1, x_2, ..., x_n),$$
$$x_2' = f_2 (t, x_1, x_2, ..., x_n), \tag{A.3}$$
$$. . . . .$$
$$x'_n = f_n (t, x_1, x_2, ..., x_n),$$

which is usually called a **normal** system of DE, where it is supposed that the number of unknowns equals the number of equations. Later, at the end of this appendix, it will be shown that notation in the form of dynamic state-space equations is in fact a matrix representation of the normal system of DE.

The solution of the above system is a set of n functions $x_1(t)$, $x_2(t)$, $x_3(t)$, ..., $x_n(t)$ that satisfy all the above equations.

A special (or particular) solution is the set of such functions that also satisfy all the initial conditions

$$x_1(t=t_0) = x_1(t_0), \; x_2(t=t_0) = x_2(t_0), \; ...., \; x_n(t=t_0) = x_n(t_0), \qquad (A.4)$$

where $x_1(t_0)$, ..., $x_n(t_0)$ are given numbers.

Here this will not be carried out, but the existence and uniqueness theorem for the solution of a normal system of DE would not be difficult to prove. We consider it more useful to show the opposite of the above derivation, i.e. that the normal system of n first-order DE is equivalent to one n-th-order DE. With this in mind, we will differentiate with respect to t the first equation from the system (A.3)

$$\frac{d^2x_1}{dt^2} = \frac{\partial f_1}{\partial t} + \frac{\partial f_1}{\partial x_1}\frac{dx_1}{dt} + \frac{\partial f_1}{\partial x_2}\frac{dx_2}{dt} + ... + \frac{\partial f_1}{\partial x_n}\frac{dx_n}{dt} \quad .$$

Replacing the expressions $dx_i/dt = x'_i$ by their equivalent functions $f_i\,(t, x_1, x_2, ..., x_n)$ yields

$$\frac{d^2x_1}{dt^2} = \frac{\partial f_1}{\partial t} + \frac{\partial f_1}{\partial x_1}f_1 + \frac{\partial f_1}{\partial x_2}f_2 + ... + \frac{\partial f_1}{\partial x_n}f_n \quad ,$$

$$\frac{d^2x_1}{dt^2} = F_2(t, x_1, x_2, ..., x_n) \qquad (A.5)$$

Differentiating the last equation and referring to (A.3) yields

$$\frac{d^3x_1}{dt^3} = \frac{\partial F_2}{\partial t} + \frac{\partial F_2}{\partial x_1}f_1 + \frac{\partial F_2}{\partial x_2}f_2 + ... + \frac{\partial F_2}{\partial x_n}f_n \quad .$$

$$\frac{d^3x_1}{dt^3} = F_3(t, x_1, x_2, ..., x_n) \qquad (A.6)$$

If this procedure is continued, we obtain

$$\frac{d^4x_1}{dt^4} = F_4(t, x_1, x_2, ..., x_n)$$

. . . .
$$\qquad (A.7)$$

$$\frac{d^n x_1}{dt^n} = F_n(t, x_1, x_2, ..., x_n) \ .$$

Now it is possible to write the following system of equations

$$\frac{dx_1}{dt} = f_1(t, x_1, x_2, ..., x_n)$$

$$\frac{d^2 x_1}{dt^2} = F_2(t, x_1, x_2, ..., x_n)$$

$$\cdots \cdots \cdots \cdots \cdots \cdots \qquad \qquad \qquad (A.8)$$

$$\frac{d^n x_1}{dt^n} = F_n(t, x_1, x_2, ..., x_n)$$

From the above system (A.8) we can express (n-1) unknown functions x_2, x_3, ..., x_n with the help of the function x_1 and its derivatives of up to, and including, the (n-1)st order. If those functions are substituted into the last equation of the above system, we finally obtain

$$\frac{d^n x_1}{dt^n} = G(t, x_1, \frac{dx_1}{dt}, ..., \frac{d^{n-1} x_1}{dt^n}) \ . \qquad \qquad (A.9)$$

To prove the validity of what has just been derived we will use the general solution of the last equation, which includes n unknown constants

$$x_1 = g_1(t, C_1, C_2, ..., C_n) \ . \qquad \qquad (A.10)$$

Substituting x_1 and all its derivatives into the expressions for x_2, x_3, ..., x_n, obtained from the first n-1 equations of the system (A.8), yields the following solution

$$x_1 = g_1(t, C_1, C_2, ..., C_n),$$
$$x_2 = g_2(t, C_1, C_2, ..., C_n), \qquad \qquad (A.11)$$
$$\cdots \cdots \cdots \cdots \cdots \cdots \cdots$$
$$x_n = g_n(t, C_1, C_2, ..., C_n).$$

This ends our proof, since it is not difficult to show that the system (A.11) in fact represents the solution of the initial normal system of first-order DE (A.3).

The above lines have shown the relationship between one n-th-order DE and a system of first-order DE, which is often used in practice to solve a dynamic problem. In that technical practice, however, it has been shown useful to distinguish between the variables, or functions, and to denote them by different letters appropriate to their physical meaning. In all the examples we showed, whose dynamics was described by one or several first-order ODE, it was always possible to distinguish between three basic variables (functions):

- input variables were denoted $u(t)$

- state variables were denoted $x(t)$

- output variables were denoted $y(t)$

Figure A.1 shows graphically the relations among these variables.

In many of the examples we pointed to the relativity of the selection of state and output variables, while the input functions were usually known and completely determined.

Nevertheless, state variables are not arbitrary variables in the process and they must be selected to correspond with their following definition:

State variables are process variables whose knowledge at the moment t_0, knowing the laws governing the behavior of the input variables for every $t > t_0$, enables the complete determination of the output variables for every $t > t_0$.

State variables are time variant (they are functions) and the set of their momentary values shows the state of the process (system, object). If it is necessary to know n state variables to describe a process, then the process is said to be of n-th order and its mathematical description is given by a system of n first-order DE.

Where, then, does the freedom of choice for state variables lie? This is the easiest to show on an example, and the simpler the example, the clearer the proof. Let us examine the case of the liquid storage tank (Section 2.1, Example 2), for which Equations (2.1-1.27) and (2.1-1.28) already showed that it is possible to select output variables freely depending on needs. In that example we selected changes in liquid level ΔH as the state variable, since changes in ΔH really did uniquely determine (through the equations we mentioned) changes in the output variables for arbitrary input variables $\Delta m_i(t)$ and $\Delta A_o(t)$. However, the state of that process, i.e. the operating regime, is, in fact, determined by the amount of liquid that is stored in the tank at the moment t_o, and the liquid level is only the measure of that amount. Thus the idea that changes in stored mass ΔM could also be selected as the state variable is not incomprehensible, especially as for this tank of constant surface area there is

input variables   state variables   output variables
(inputs, excitations,   (outputs, responses,
disturbances, noises)   solutions)

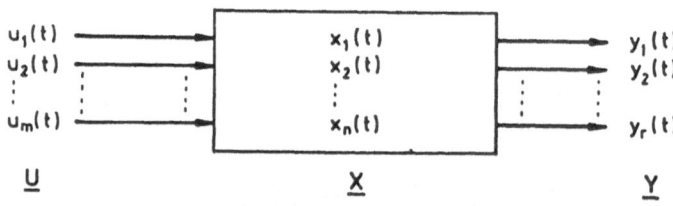

Fig. A.1 Illustration of the definition of input, state and output variables

$$\Delta M = A\rho\Delta H \quad . \tag{A.12}$$

If ΔH is expressed from (A.12), the state equation changes appearance and becomes

$$\left[\Delta\dot{M}\right] = \left[-\frac{1}{T}\right]\left[\Delta M\right] + \left[1 \quad -K_A\right]\begin{bmatrix}\Delta m_i \\ \Delta A_o\end{bmatrix} \quad . \tag{A.13}$$

And, of course, if liquid level is a controlled variable, the output equation now also changes appearance and instead of (2.1-1.27) becomes

$$\left[\Delta H\right] = \left[\frac{1}{A\rho}\right]\left[\Delta M\right] + \begin{bmatrix}0 & 0\end{bmatrix}\begin{bmatrix}\Delta m_i \\ \Delta A_o\end{bmatrix} \quad . \tag{A.14}$$

It is very important to note that the order of the system is uniquely determined, and whichever variable is selected, the order of the system cannot and must not change.

It must also be mentioned that the state variables need not have a clear logical or physical meaning, although it is in practice useful to chose them so that they do have a clear meaning or, even, so that they are always measurable and accessible variables in the process. (The demand for them to be measurable and accessible can often not be fulfilled.) Thus, for example, multiplying the whole state equation (2.1-1.24) by n and introducing a new state variable $\Delta V = n\Delta H$ (which equals an n-fold change in liquid level) yields

$$\left[\Delta V\right]' = \left[-\frac{1}{T}\right]\left[\Delta V\right] + \left[\frac{n}{C} \quad -\frac{K_A n}{C}\right]\begin{bmatrix}\Delta m_i \\ \Delta A_o\end{bmatrix}. \tag{A.15}$$

Here we can conjecture about the physical meaning of the state variable ΔV, but it is clear that the selection is neither logical nor natural. Nevertheless, solving (A.15) would give completely valid solutions for ΔV, with whose help it would be possible to determine the other variables of interest as well.

It is also advantageous to treat the concept of **state space**. This term has come into process dynamics analysis from mathematics and it has a clear physical meaning. Turning again to the same example of the storage tank whose model is given by the state Equation (2.1-1.24) and the output Equation (2.1-1.27), we can repeat that it is a first-order system since it consists of only one mass storage tank. Therefore, as time passes only one variable changes due to input variable change, and that is the liquid level ΔH. (Obviously, output flow Δm_o also changes, but that change depends both on the input variables $u(t)$ and ΔH, which means on that state variable, while change in ΔH depends only on itself and on input variables.) The process is **one-dimensional** in the sense of there being only **one dependent variable**. Changes in that state variable can be shown in two ways: in a $t - \Delta H$ diagram and in the one-dimensional ΔH space in which time t is given parametrically. That space is called **the space of state changes** or **state space** and is shown in Figure A.2b for the case when at $t_0 = 0$ water flow into the tank is discontinued and water level in the tank equals 1 m, which results in free outflow. (Since this is a one-dimensional system, we could also have called it a line of state.)

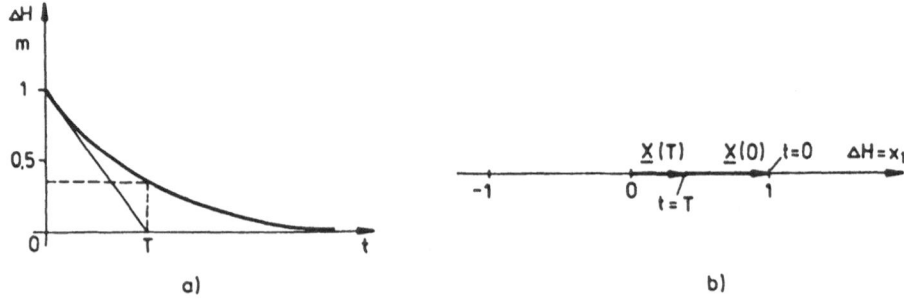

Fig A.2 Graphical presentation of state changes in a
one-dimensional problem - line of state

If the variable is stored in two tanks the process is of second order
and there are two state variables.

In Section 2.2-3 (Example 1, Figure 2.2-3.2) the t-x diagram shows how
those two state variables change (x_1 = ΔP_2 and x_2 = Δm_1) and Figure A.3
shows the same in the state space (plane), i.e. in a ΔP_2 - Δm_1 diagram
where time t is the parameter. The trajectory of state change is typical for
periodic damped processes in which the state variables, for a given input
Δm_2 = 1, converge to a new steady state (0,1).

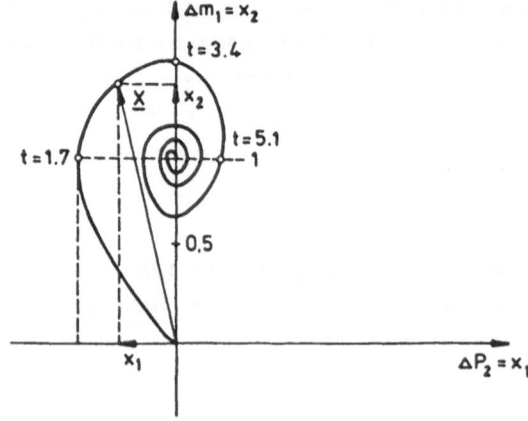

Fig. A.3 Graphical presentation of state changes in a two-dimensional
problem - plane of state, phase plane

In Figure A.3, as indeed in all the other presentations of state space, it must be observed that only one trajectory passes through one point of the state space. This is a direct consequence of the definition of the state variables themselves, according to which, for given input values $u_j(t)$ ($j = 1, 2, ..., m$), they determine completely and uniquely the future behavior of the system for every $t > t_0$. (Exceptions are so-called singular points in state space.)

Finally, in the case of three storage tanks in series the state space would be three-dimensional and the trajectory could be shown as a curve in three-dimensional space with coordinates x_1 - x_2 - x_3 (ΔH_1 - ΔH_2 - ΔH_3, in the example with three storage tanks - Section 2.5).

Therefore, we can say that the dimension of the state space equals the number of state variables necessary to describe the dynamics of the process under observation. If that dimension is $n < 3$ the state space can also be represented (graphically), but for mathematical processing the impossibility of graphical representation for cases of $n > 3$ is of no essential meaning. (In mechanics the state trajectory in second-order systems is very frequently represented, and then the state space (plane) is called a **phase plane**.)

It must also be emphasized that in a large number of cases the order or dimension of the process model, i.e. the number of first-order DE that describe its dynamics, is not firmly determined but greatly depends on the final purpose of the modeling. Or, in other words, almost every process can be described by models of different order-from those that describe only steady states (0-th order models) to models in the form of partial differential equations (infinite-order models). Therefore, it must always be remembered that in most cases a dynamics process model is in fact a hierarchy of models describing that process - from very simple ones to those of great complexity. The final goal of modeling, the equipment that is to be used, the mathematical methods, procedures and programs that have been developed, economic considerations, time and personnel - all these are factors that determine the complexity, accuracy and price of the model derived.

Finally, before presenting the most general system of first-order DE in the form of matrix DE of state space, it will be useful to give a clear and simple example of a model using the notation u, x and y for the variables of input, state and output. That transition to the general and abstract form of notation will be shown on Example 1 from Section 2.1-2 on gas storage tanks, whose model is given by Equations (2.1-2.10) and (2.1-2.9). We introduce the notation:

$$x_1(t) = P(t) \qquad \text{... the state variable is the pressure in the tank}$$

$$u_1(t) = m_i(t)$$
$$u_2(t) = P_o(t) \qquad \text{... input variables}$$
$$u_3(t) = A_o(t)$$

$$y_1(t) = P(t) \qquad \text{... output variables}$$
$$y_2(t) = m_o(t)$$

In further equations we will no longer emphasize that all these are time-dependent variables, so the dynamic model of the process of filling a storage tank becomes

$$
\begin{aligned}
x_1' &= -k_1 u_3 \sqrt{u_2(x_1 - u_2)} + u_1 \quad , \\
y_1 &= x_1 + 0 \cdot u_1 + 0 \cdot u_2 + 0 \cdot u_3 \quad , \\
y_2 &= k_1 u_3 \sqrt{u_2(x_1 - u_2)} + 0 \cdot u_1 \quad .
\end{aligned}
\qquad (A.16)
$$

In this system the dependencies that do not exist are especially emphasized (by multiplication with zero). In general, therefore, the above system has the following functional form

$$
\begin{aligned}
x_1' &= f_1 (x_1, u_1, u_2, u_3), \quad NL \\
y_1 &= g_1 (x_1, u_1, u_2, u_3), \quad L \\
y_2 &= g_2 (x_1, u_1, u_2, u_3), \quad NL
\end{aligned}
\qquad (A.17)
$$

The original model given by (A.17) is nonlinear and in principle further analysis of the process can take two different directions. The first is direct numerical integration on a computer, which presents no special difficulties. The second has been shown in Section 2.1-2; it includes linearizing the original model, translating the linear model into matrix state (space) equation notation and obtaining a matrix transfer function which clearly shows the character of the output variables' dynamic dependence on the input variables. Without repeating the procedure of linearization, it is useful here to give a more abstract form of the linear model shown by the state-space equations (2.1-2.26) and (2.1-2.27)

$$x'_1 = a_{11}x_1 + b_{11}u_1 + b_{12}u_2 + b_{13}u_3,$$
$$y_1 = 1 \cdot x_1 + 0 \cdot u_1 + 0 \cdot u_2 + 0 \cdot u_3, \qquad \text{(A.18)}$$
$$y_2 = c_{21}x_1 + 0 \cdot u_1 + b_{22} \cdot u_2 + b_{23} \cdot u_3.$$

It must be noticed that the zeros in the linearized model appear in the same places where there was no dependence in the nonlinear model.

Now it is possible to give the most general presentation of the dynamic model of a process which has m inputs, n states (the system is of n-th order) and r outputs

$$x'_1 = f_1 (x_1, ..., x_n, u_1, ..., u_m),$$
$$x'_2 = f_2 (x_1, ..., x_n, u_1, ..., u_m),$$
$$\cdots \cdots \cdots \cdots \cdots \cdots \cdots \qquad \text{(A.19)}$$
$$x'_n = f_n (x_1, ..., x_n, u_1, ..., u_m).$$

In the general case, the functions f_i $(i = 1, 2, ..., n)$ are nonlinear. The output variables y_j $(j = 1, 2, ..., r)$ are related to the state variables and the input variables by static, algebraic equations of this general form

$$y_1 = g_1 (x_1, x_2, ..., x_n, u_1, u_2, ..., u_m),$$
$$y_2 = g_2 (x_1, x_2, ..., x_n, u_1, u_2, ..., u_m),$$
$$\cdots \cdots \cdots \cdots \cdots \cdots \cdots \cdots \qquad \text{(A.20)}$$
$$y_r = g_r (x_1, x_2, ..., x_n, u_1, u_2, ..., u_m).$$

Functions g_j $(j = 1, 2, ..., r)$ are also in general nonlinear. Equations (A.16) are an example of one such nonlinear system in which the nonlinearities are of the square root and product of functions $x(t)$ and $u(t)$ type. In most of this book we made efforts to transform originally nonlinear models given by (A.19) and (A.20) into a linear form of the following kind

$$x'_1 = a_{11}x_1 + a_{12}x_2 + ... + a_{1n}x_n + b_{11}u_1 + b_{12}u_2 + ... + b_{1m}u_m,$$
$$x'_2 = a_{21}x_1 + a_{22}x_2 + ... + a_{2n}x_n + b_{21}u_1 + b_{22}u_2 + ... + b_{2m}u_m, \quad \text{(A.21)}$$
$$\cdots \cdots \cdots \cdots \cdots \cdots \cdots \cdots \cdots \cdots \cdots$$
$$x'_n = a_{n1}x_1 + a_{n2}x_2 + ... + a_{nn}x_n + b_{n1}u_1 + b_{n2}u_2 + ... + b_{nm}u_m.$$

$$y_1 = c_{11}x_1 + c_{12}x_2 + ... + c_{1n}x_n + d_{11}u_1 + d_{12}u_2 + ... + d_{1m}u_m,$$
$$y_2 = c_{21}x_1 + c_{22}x_2 + ... + c_{2n}x_n + d_{21}u_1 + d_{22}u_2 + ... + d_{2m}u_m, \quad \text{(A.22)}$$
$$\cdots \cdots \cdots \cdots \cdots \cdots \cdots \cdots \cdots \cdots \cdots$$
$$y_r = c_{r1}x_1 + c_{r2}x_2 + ... + c_{rn}x_n + d_{r1}u_1 + d_{r2}u_2 + ... + d_{rm}u_m.$$

In vector notation the above equations get the well-known form of matrix state-space equations

$$X' = A X + B U \; ,$$
$$Y = C X + D U \; . \tag{A.23}$$

Now it is easy to see that what was shown earlier about the selection of state variables not being unique is also valid for the most general case given by model (A.23). It is known that the coordinates of a point in n-dimensional space depend on the selection of the base, i.e. on the system of mutually perpendicular unit vectors along which the coordinate axes are directed. By changing the base we can move from the state vector X to the state vector $\dot{x}$ using the simple transformation

$$\dot{x} = MX, \; M \neq 0 \; . \tag{A.24}$$

M is an nonsingular matrix of order (nxn). For every such transformation there is also an inverse transformation

$$X = M^{-1} \dot{x} \; . \tag{A.25}$$

If, thus, the system of state equations (A.23) is valid for the vector of state variables X, then for the new **vector of state variables** $\dot{x}$ there will be

$$\dot{x}' = \dot{A} \dot{x} + \dot{B} U \; ,$$
$$Y = \dot{C} \dot{x} + D U \; . \tag{A.26}$$

The old and the new matrices are related as follows

$$\dot{A} = M A M^{-1} \; .$$

$$\dot{B} = M B \; . \tag{A.27}$$

$$\dot{C} = C M^{-1} \; .$$

It is obvious that there is an unlimited number of different transformation matrices M and it would be difficult to expect the new state variables to retain their originally clear physical meaning after an arbitrary transformation. Therefore, if such transformations are performed, they usually have a deliberately selected matrix **M** which serves to transform matrix **A** into a form that is suitable for the dynamic analysis of the process described, and which makes it very much easier.

Finally, we must repeat that spatially distributed processes, whose model is a system of partial differential equations (PDE), cannot (without additional procedures of discretization) directly be described by a model in the form of matrix state-space equations. This results from the fact that such a system of PDE (or just one) describes changes of process variables in every point in space inside which the process occurs, and there are, of course, infinitely many such points. Therefore, system matrix **A** would be of infinite order. Only after spatial discretization has been performed can a (reduced) model of the spatially distributed process be obtained in the form of a system of ODE like the one given by (A.19) and (A.20), from which it is not far to state-space form (A.23).

It must also be remarked that, although such examples were not treated in this book, the coefficients of matrices **A**, **B**, **C**, and **D** as a rule need not be constants, but can be functions of time t. Such processes will be called **time variant processes**, and to distinguish them from the notation used up to now, time t will appear explicitly in the functions f_i (i = 1, 2, ..., n) and g_j (j = 1, 2, ..., r), as was the case in (A.1). A time-variant system would, thus, be characterized by the following notation

$$x'_i = f_i(t, x_1, x_2, ..., x_n, u_1, u_2, ..., u_m) \ (i = 1, n) \tag{A.28}$$

To get a solution for such systems it is necessary, besides knowing the initial conditions x_i (i = 1, 2, ..., n) and the input functions $u_j(t > t_0)$(j = 1, 2, ..., m), to know also the time t when the disturbance occurs and calculation begins, because now the law governing state variable changes will no longer be the same at two different moments for the same initial conditions and inputs.

MODEL LINEARIZATION

Linearization is the replacement of nonlinear mathematical expressions relating process variables with linear ones. Linearization has been used in this book in almost every example, which shows that **in everyday technical practice originally linear models are in most cases the exception** and that in by far the greatest number of processes the relations between the input, state and output variables are nonlinear. The persistence, however, that is shown in deriving models and transforming them into linear forms points to the great theoretical and practical value of such linear mathematical forms.

The main reasons that justify this procedure are the following:

> - for small deviations from the usual operating states the linear model behaves like the original nonlinear process,

> - linear models are solved using the highly-developed and powerful mathematical apparatus of linear algebra, which is today part of standard mathematical computer libraries,

> - the principle of superposition is used, so it is sufficient to examine systems with unique and simple input signals on the basis of whose responses dynamic properties can be typified (into proportional, integral, derivative), which leads to a generalized and unified approach to the analysis of dynamics of otherwise different processes,

> - as stability is a property of the process itself and does not depend on the initial conditions or disturbances, the examination of the stability of a certain steady state or operating regime is reduced to an analysis of the eigenvalues of system matrix $\mathbf{A}$.

None of these properties (except, of course, the first) characterize nonlinear models, whose basic feature is that general solutions, methods and approaches cannot be used. Every nonlinear process contains within it a different form of nonlinearity that demands a special method of solution. This impossibility of a unified approach (the impossibility of generalization) is one of the reasons why nonlinear theory is still in the phase of development and intense study. Nevertheless, one unified approach to such systems does exist, and that is the procedure of numerical simulation of the nonlinear model. Today this is the most frequently used method for analyzing such processes and it can be applied without difficulty to every example in this book.

In practice the system variables in devices, plants and processes are connected by various forms of nonlinear relations: limiters, switches, breakers, flip-flops, saturations, dead zones, friction, clearances and hystereses are only a small number of typical nonlinear relations. In the text of this book we encountered nonlinearities that were analytically expressed in the form of products between variables, square roots, squares and the like, or nonlinear relations were shown graphically in diagrams of characteristics.

For such varying cases of nonlinearity, and depending on the final purpose of the model, today the following developed methods of linearization are usually used:

> - the method of tangential approximation or linearization about the operating (usually steady) state, in which nonlinear characteristics are replaced by linear ones (in two-dimensional space this means replacing the curve by the tangent in the operating point). In the mathematical sense this means developing the functions into a Taylor series in which all the higher (nonlinear) terms are discarded. The procedure is the more successful, the smaller the deviations from the operating state.

> - the method of the descriptive function or harmonic linearization, which linearizes processes in the frequency domain and not in the time domain, as in the above case. In control theory this method is often and successfully used and its main feature is that only the signal of the basic frequency is recorded in analysis while the higher harmonics are considered damped. The success of the method depends on how similar the processes are to low-frequency filters.

> - the method of statistical linearization is also performed in the

time domain. Here the nonlinear process is replaced by an equivalent linear model assuming that the disturbance has a normal (Gaussean) distribution,

- the combined method of harmonic and statistical linearization.

In this book deterministic processes were analyzed in the time domain (smaller excursions into the domain of the complex variable were performed to obtain and demonstrate transfer functions), so the subsequent lines will show only the method of tangential approximation because it was used in all the examples in the preceding sections and is today still the most frequently used. (In Anglo-American literature it is often called perturbance analysis. This name is used to show that the method yields linear models that describe small deviations, perturbances, from the operating state.)

The basic assumption that must be fulfilled if we want to apply this method is that the relations between variables must be given in a functional form that has a finite first derivative in the operating point about which linearization is performed (i.e. that they are differentiable at least once), and if the relations are shown by a characteristic curve, then it must have a clearly determined tangent in that operating point. This method, is, as a rule, not suitable for step and other broken or discontinuous nonlinear relations.

Tangential approximation of a nonlinear function is performed by expansion into a Taylor series (if the function is given in analytical form) or graphically if the relations are shown in diagram. This second method was shown in detail in Section 2.2 where the statical characteristics of the centrifugal pump and of the regulation valve were linearized. Using Figures 2.2-1.2 and 2.2-1.3, we obtained linear expressions for the pump, (2.2-1.10) and (2.2-1.16), and for the valve, (2.2-1.11) and (2.2-1.17). The same procedure can be used in all the other cases and here it will not be repeated. Before the method of tangential approximation is demonstrated, the state and output equations given by (A.19) and (A.20) should also be given in vector notation

$$\mathbf{X}' = \mathbf{f}(\mathbf{X}, \mathbf{U}) \tag{A.29}$$

$$\mathbf{Y} = \mathbf{g}(\mathbf{X}, \mathbf{U}) \tag{A.30}$$

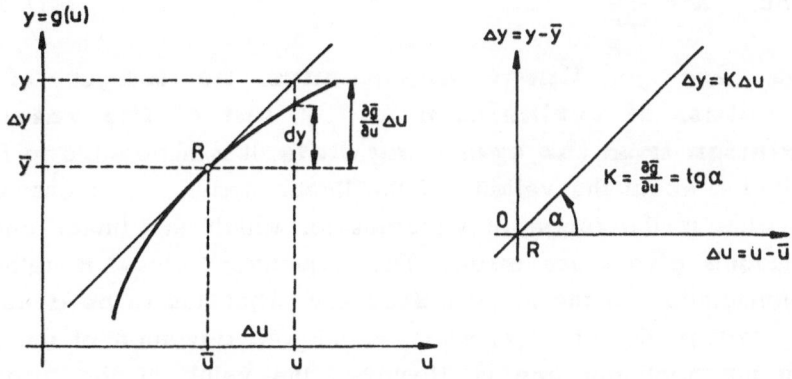

Fig. A.4 The meaning of tangential approximation

The meaning of linearization is the easiest to comprehend in two-dimensional space where relations among variables can be shown by a curve and a tangent in the operating point. In the case of three-dimensional relations the graphical presentation is a surface in this space and the plane tangential to it, and in the most general case it is a hypersurface and its tangential hyperplane, which cannot be shown graphically. Let the output variable, therefore, be related only with the input variable by the following relation

$$y = g(u) \ . \tag{A.31}$$

For a given $\bar{u}$, $\bar{y}$ is determined and the operating point $R(\bar{u}, \bar{y})$ is shown on Figure A.4

Expansion into a Taylor series about the point $R(\bar{u}, \bar{y})$ yields

$$y = \bar{y} + \frac{d\bar{g}}{du}(u - \bar{u}) + \frac{1}{2!}\frac{d^2\bar{g}}{du^2}(u - \bar{u})^2 + \ldots \tag{A.32}$$

The deviations shown here are calculated in the operating point R. If analysis is limited only to small deviations, then the square term and all the following terms of higher order can be neglected because they are much smaller than the linear term in the expansion, so we get

$$y - \bar{y} = \frac{d\bar{g}}{du}(u - \bar{u}) \ . \tag{A.33}$$

$$\Delta y = K\Delta u, \quad K = \frac{d\bar{g}}{du} \quad . \tag{A.34}$$

Now we have got **linear dependance, no longer of the absolute values of variables u and y, but of the values of their deviation from the operating state**. It is impossible to form a general criterion about the validity of the linear model, i.e. to answer the question - what is the range of variables for which the linear and the nonlinear models give close results. The adjective " close" is rather an unprecise formulation for the model's accuracy. What has to be done is to check, for every particular case, what the allowed deviation of the linear model from the nonlinear one is. However, the value of the error can always be obtained because it equals the magnitude of the discarded higher-order terms. The following simple example shows the difference in results yielded by the nonlinear and the linear models.

$$y = u^2, \quad R(2, 4) \qquad \bar{u} = 2, \bar{y} = 4$$

Expansion into a series gives

$$y = \bar{y} + 2\bar{u}\cdot(u - \bar{u}) + \frac{1}{2!}\cdot 2(u - \bar{u})^2$$

$$y = \bar{y} + 4(u - \bar{u}) + (u - \bar{u})^2 \qquad\qquad \text{nonlinear model}$$

If the last, square term in the expansion is neglected, the following linear model is obtained

$$y - \bar{y} = 4(u - \bar{u})$$

$$\Delta y = 4\Delta u \qquad\qquad\qquad \text{linear model}$$

Obviously, the difference between those two models, or the error made by neglecting higher-order terms, equals

$$e = y_{NL} - y_L = (u - \bar{u})^2 = \Delta u^2$$

Table showing the results and their graphical presentation

Nonlinear model				Linear model				Error e
u	y_{NL}	Δu	Δy	Δu	Δy	u	y_L	$e = y_{NL} - y_L = \Delta u^2$
0	0	-2	-4	-2	-8	0	-4	4
1	1	-1	-3	-1	-4	1	0	1
2	4	0	0	0	0	2	4	0
3	9	1	5	1	4	3	8	1
4	16	2	12	2	8	4	12	4

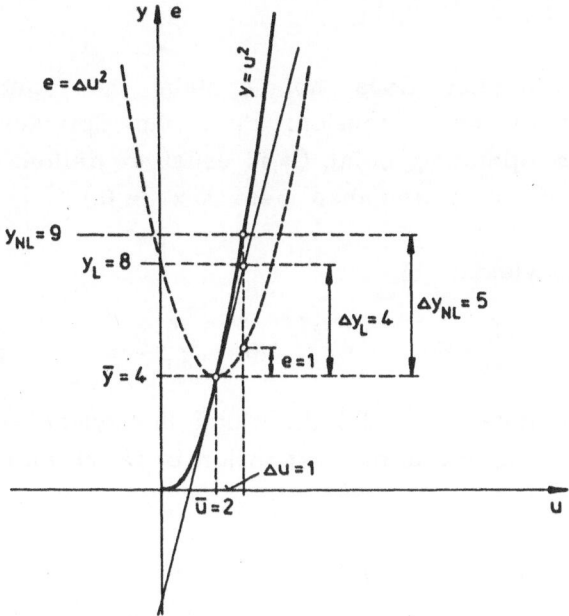

Fig. A.5 Error due to the linearization of the curve u^2

The figure shows the value of the error. For example, if the input variable u deviates by 50%, i.e. for $\Delta u = 1$, the linear model makes an error of 11.1%, or the value given by the linear model is 11.1% smaller than the real value given by the nonlinear model. Here we must immediately consider whether there are such processes in which the input variable changes its value by 50%, and whether it matters if the error in calculation is of the above order of magnitude. For smaller deviations the errors are smaller, almost negligible.

The same procedure that has been shown for the two-dimensional problem can be used when y is a function of several input and state variables, i.e. when it is given in the form (A.20) or (A.30). In such higher-order processes, in which we distinguish between state and input variables, expansion into a Taylor series will be shown on the i-th differential state equation from the system (A.19)

$$x_i' = f_i(x_1, ..., x_n, u_1, ..., u_m) \tag{A.35}$$

$$x_i' = \bar{x}_i' + \sum_{j=1}^{n} \frac{\partial f_i}{\partial x_j}(x_j - \bar{x}_j) + \sum_{k=1}^{m} \frac{\partial f_i}{\partial u_k}(u_k - \bar{u}_k) + ... \tag{A.36}$$

$$\left[\bar{x}_i' = f_i(\bar{x}_1, ..., \bar{x}_n, \bar{u}_1, ..., \bar{u}_m) \right]$$

The above equation does not contain the square and other higher-order term of the expansion. $\bar{x}'_i$ is the derivative of the state variable x_i in the operating point. (It is usual to perform linearization of DE about the steady state, and then there is $\bar{x}'_i = 0$.)

Equation (A.36) yields

$$x_i' - \bar{x}_i' = \Delta x_i' = \frac{\partial f_i}{\partial x_1} \Delta x_1 + ... + \frac{\partial f_i}{\partial x_n} \Delta x_n + \frac{\partial f_i}{\partial u_1} \Delta u_1 + ... + \frac{\partial f_i}{\partial u_m} \Delta u_m \tag{A.37}$$

The system of equations (A.21) (in which the symbol Δ has been left out) was obtained from the above expansion of (A.19), and the coefficients a and b are

$$a_{ij} = \frac{\partial f_i}{\partial x_j} \; , \qquad b_{ik} = \frac{\partial f_i}{\partial u_k} \quad i, j = 1, ..., n \qquad k = 1, ..., m$$

The value of the partial derivatives is determined in the operating state observed.

The coefficients a_{ij} and b_{ik} are members of the matrices **A** and **B**, so we can write

$$\mathbf{A} = \left(\frac{\partial \mathbf{f}}{\partial \mathbf{X}} \right)_R \; , \qquad \mathbf{B} = \left(\frac{\partial \mathbf{f}}{\partial \mathbf{U}} \right)_R \tag{A.39}$$

The subscript R denotes that the partial derivatives are calculated in the operating state R.

Exactly the same procedure is used to linearize the output equations (A.20) and (A.30), only instead of x_i and the function f_i we now have output variables y_i and the function g_i.

Tangential approximation is a method that can be widely applied, but it must always be born in mind that the domain in which the variables can change after linearization has been performed is limited to 5, 10 or 20% of the steady values of those variables. The following figure is a graphical presentation of how two completely different processes can have the same linear model for a given ū.

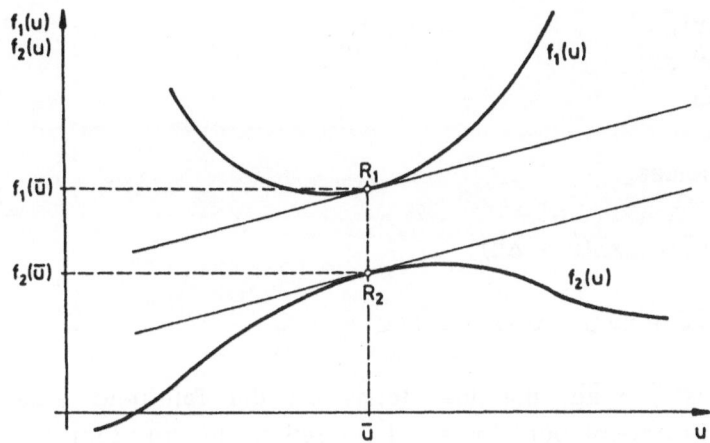

Fig. A.6 Two different processes with the same linear model

It must also be said that in the case of a nonlinear model it is very important to determine the operating state in which linearization was performed. The nonlinear model can have one, two or more steady states, or it can have none, so we must consider problems of determining the steady state as well. Here we will not enter this analysis and the reader is referred, as a small reminder, to the last example of Section 2.4, where it was shown that the moving pendulum has two steady states in which linearization gives different matrices **A**, which also means different linear models with essentially different dynamic properties.

In practice, other ways for obtaining a tangential approximation are also often used. The most widespread procedure is to substitute into nonlinear expressions the values for variables which are represented as the sum of the variable showing the steady state and that showing deviation from it. On the next simple example we will show this method, which avoids partial differentiation.

Linearize the following nonlinear output equation

$$y = xu \quad , \tag{A.40}$$

about its steady state $(\overline{x}, \overline{u}, \overline{y})$.

Introducing

$$\begin{aligned} y &= \overline{y} + \Delta y \ , \\ x &= \overline{x} + \Delta x \ , \\ u &= \overline{u} + \Delta u \ , \end{aligned} \tag{A.41}$$

(A.40) becomes

$$\overline{y} + \Delta y = (\overline{x} + \Delta x) \cdot (\overline{u} + \Delta u)$$

$$\overline{y} + \Delta y = \overline{xu} + \overline{x}\Delta u + \overline{u}\Delta x + \Delta x \Delta u \tag{A.42}$$

As there is $\overline{y} = \overline{xu}$, the first terms on the left-hand side and the right-hand side cancel out. For small deviations the product $\Delta x \Delta u$ is very small, and can be neglected. Thus (A.42) yields the following linear expression

$$\Delta y = \overline{u}\Delta x + \overline{x}\Delta u \quad . \tag{A.43}$$

Of course, the same result would have been obtained by expansion into the Taylor series

$$y = \overline{y} + \left(\frac{\partial y}{\partial x}\right)_R (x - \overline{x}) + \left(\frac{\partial y}{\partial u}\right)_R (u - \overline{u}) \quad . \tag{A.44}$$

$$\left(\frac{\partial y}{\partial x}\right)_R = \overline{u} \ , \qquad \left(\frac{\partial y}{\partial u}\right)_R = \overline{x} \ ,$$

$$\Delta y = \overline{u}\Delta x + \overline{x}\Delta u \quad . \tag{A.45}$$

There is one more method which indirectly represents expansion into a Taylor series in which higher-order terms have been neglected. It is the **method of differentiating** the whole nonlinear expression, after which infinitesimal changes are replaced by finite deviations. On the same example this method gives

$$dy = (\frac{\partial y}{\partial x})_R dx + (\frac{\partial y}{\partial u})_R du \quad .$$

$$dy = \bar{u}dx + \bar{x}du \quad . \tag{A.46}$$

Now the symbols for infinitesimal change d are replaced by the symbols of finite deviation Δ and the same linear model as before is obtained

$$\Delta y = \bar{u}\Delta x + \bar{x}\Delta u \quad .$$

This method of differentiating is probably the most suitable for use. It has been used in almost all the examples in this book and was also used to linearize the PDE in Section 3.3 - see the derivation given by Equations (3.3-67) - (3.3-71).

There are other methods for obtaining linear models also. One of them was shown in the last example of Section 2.4, where the linear model of the moving pendulum was reached after so-called algebraic linearization - the substitution of sine and cosine functions by functions similar to them for small angular displacements. Besides this method, that of least squares is used in practice to determine coefficients of linear development. Today, thanks primarily to possibilities opened up by computers and in cases when it is very difficult (because of the size and complexity of the model) to perform differentiation analytically, the linear model is arrived at numerically.

SELECT FUNCTIONS OF COMPLEX VARIABLE

$s = o + j\omega = x + jy$

$e^s = 1 + \dfrac{s}{1!} + \dfrac{s^2}{2!} + ... + \dfrac{s^n}{n!} + ... = \displaystyle\sum_{k=0}^{n} \dfrac{s^k}{k!}$

$e^s = e^x(\cos y + j\sin y)$ $\qquad\qquad$ $e^{js} = \cos s + j\sin s$

$\cos s = \dfrac{e^{js} + e^{-js}}{2}$ $\qquad\qquad$ $\sin s = \dfrac{e^{js} - e^{-js}}{2j}$

$\cos s = 1 - \dfrac{s^2}{2!} + \dfrac{s^4}{4!} - ... + (-1)^n \dfrac{s^{2n}}{(2n)!} + ...$

$\cos s = \cos x \cdot \text{ch} y - j\sin x \cdot \text{sh} y$

$\sin s = s - \dfrac{s^3}{3!} + \dfrac{s^5}{5!} - ... + (-1)^n \dfrac{s^{2n+1}}{(2n+1)!} + ...$

$\sin s = \sin x \cdot \text{ch} y + j\cos x \cdot \text{sh} y$

$\text{ch}\, s = 1 + \dfrac{s^2}{2!} + \dfrac{s^4}{4!} + ... + \dfrac{s^{2n}}{(2n)!} + ... = \dfrac{e^s + e^{-s}}{2}$

$\text{ch}\, s = \text{ch} x \cdot \cos y + j\text{sh} x \cdot \sin y$

$\text{sh}\, s = s + \dfrac{s^3}{3!} + \dfrac{s^5}{5!} + ... + \dfrac{s^{2n+1}}{(2n+1)!} + ... = \dfrac{e^s - e^{-s}}{2}$

$\text{sh}\, s = \text{sh} x \cdot \cos y + j\text{ch} x \cdot \sin y$

$\text{th}\, s = \dfrac{\text{sh}\, s}{\text{ch}\, s}$ $\qquad\qquad\qquad$ $\text{cth}\, s = \dfrac{\text{ch}\, s}{\text{sh}\, s}$

BASIC DYNAMIC PROPERTIES

Type	Differential equation	Transfer function	Transient function [response to u=h=1(t≥0)]	Poles x Zeros o
P	$x = K_p u$	K_p	K_p	
P_1	$T\dfrac{dx}{dt} + x = K_p u$	$\dfrac{K_p}{Ts+1}$	$K_p(1 - e^{-\frac{t}{T}})$	
P_2	periodic case $T_2^2\dfrac{d^2x}{dt^2} + T\dfrac{dx}{dt} + x = K_p u$ $\omega_n = 1/T_2,\ \xi = T/2T_2$	$\dfrac{K_p}{T_2^2 s^2 + Ts + 1}$ $\omega_p = \omega_n\sqrt{1-\xi^2}$ $\varphi = \text{arc tg}\dfrac{\omega_p}{\omega_n\xi}$	$K_p\left[1 - \dfrac{1}{\sqrt{1-\xi^2}}e^{-\xi\omega_n t}\cdot \sin(\omega_p t + \varphi)\right]$	
I	$x = \dfrac{1}{T_i}\int_0^t u\,dt$	$\dfrac{1}{T_i s}$	$\dfrac{1}{T_i}t$	
I_1	$T\dfrac{dx}{dt} + x = \dfrac{1}{T_i}\int_0^t u\,dt$	$\dfrac{1}{T_i s(Ts+1)}$	$\dfrac{1}{T_i}[t - T(1 - e^{-\frac{t}{T}})]$	
D	$x = T_d\dfrac{du}{dt}$	$T_d s$	$T_d\delta(t)$	
D_1	$T\dfrac{dx}{dt} + x = T_d\dfrac{du}{dt}$	$\dfrac{T_d s}{Ts+1}$	$\dfrac{T_d}{T}e^{-\frac{t}{T}}$	
PD	$x = u + T_d\dfrac{du}{dt}$	$1 + T_d s$	$1 + T_d\delta(t)$	
PD_1	$T\dfrac{dx}{dt} + x = u + T_d\dfrac{du}{dt}$	$\dfrac{T_d s+1}{Ts+1}$	$1 + (\dfrac{T_d}{T} - 1)e^{-\frac{t}{T}}$	
T	$x = u(t - T_t)$	$e^{-T_t s}$	$1(t - T_t)$	
non-minimum phase	$T\dfrac{dx}{dt} + x = u - T\dfrac{du}{dt}$	$\dfrac{1 - Ts}{1 + Ts}$	$1 - 2e^{-\frac{t}{T}}$	
PI	$x = Ku + K_i\int u\,dt$	$K + \dfrac{K_i}{s}$ or $K_r(1 + \dfrac{1}{T_i s}),\ K_i = \dfrac{K_r}{T_i}$	$K_r(1 + \dfrac{t}{T_i})$	
PI_1	$T_1 x' + x = Ku + K_i\int u\,dt$	$\dfrac{K_r(1 + \dfrac{1}{T_i s})}{1 + T_1 s},\ K_i = \dfrac{K_r}{T_i}$	$K_r[\dfrac{t}{T_i} + (1 - \dfrac{T_1}{T_i})(1 - e^{-t/T_1})]$	
PID	$x = Ku + K_i\int u\,dt + K_d u'$	$K_r(1 + \dfrac{1}{T_i s} + T_d s)$ $K = K_r, K_i = \dfrac{K_r}{T_i}, K_d = K_r T_d$	$K_r[1 + \dfrac{t}{T_i} + T_d\delta(t)]$	

LIST OF SYMBOLS

(Numbers in parentheses indicate sections)

A	m^2	area
A		system's matrix in state equation
A(t)		acceleration
AV(t)		accumulated (stored) variable
a	m^2	area (2.5)
a	m/s^2	acceleration
a		arbitrary coefficient
B		input's matrix in state equation
b_v		valve coefficient (2.2-1)
C		coefficient of capacity
C		output's matrix in output equation
c	J/kgK	specific heat
c_s, c	m/s	velocity of sound (2.2-3), (3.3)
c	mol/m^3	concentration
c	N/m	spring constant
c_t	Nm/rad	torsional spring constant
D	m	diameter
D	Ns/m	viscous friction coefficient
D		input's matrix in output equation
D_t	Nms/rad	torsional viscous friction coefficient
d	m	diameter
E	J	energy
E	N/m^2	modulus of elasticity, Young's modulus (3,3)
E(t)		effort (potential)
e	J/s	energy (heat) flow rate
F	N	force (2.4)
F(t)		flow
f		various function
G(s)		transfer function
G(s)		matrix transfer function
g		various function
H	m	height of liquid level, head
h(t)		Heaviside function of unit step

I		coefficient of inertia
I_m = Mw	kgm/s	momentum
i	J/kg	enthalpy
J	kgm^2	moment of inertia
K		gain coefficient
K_v		valve characteristic
k = 1/T	1/s	heat (thermal) coefficient (3.1)
L	m	length
L_z	kgm^2/s	angular momentum
l	m	length
M	kg	mass
M_z	Nm	torque
m	kg/s	mass flow rate
N	mol	quantity of mater
n		polytropic exponent
n	s^{-1}	number of revolution
n		zeros of transfer function
n	mol/s	molal flow rate
n_d	mol/s	diffusiv molal flow rate
P	N/m^2	pressure
p(z)		initial conditions
Q, q	J/s	heat flow rate (3.2)
q	J/m^{2}s	heat flux per unit area (2.3)
R		coefficient of resistance
R	m	radius
R	J/kgK	gas constant
r	m	radius
r(t)		boundary conditions
r(P)	J/kg	specific heat of vaporization
s = σ + jω		complex variable of Laplace transformation
s	J/kgK	entropy
T	s	time constant
t	s	time coordinate
U	J	internal energy (3.1)
U	m	circumference
U		input vector
u	J/s	heat flow rate (3.1)
u	J/kg	internal energy
u_i		input variables, elements of **U**
V	m^3	volume

v	m^3/kg	specific volume
w	m/s	velocity
X		state vector
x_i		state variables, elements of **X**
Y		output vector
y_i		output variables, elements of **Y**
Z		impedance
z	m	spatial coordinate

SUBSCRIPTS

a	approximative
c	pipe
d	damping (2.4)
el	electrical
ex	external
f	fluid
i	input
in	internal (2.1-1)
k	critical
m	mass (2.4)
n	natural (2.4)
o	output
o	spring
p	pump
s	sound (2.2-3), (3.3)
t	torsional (2.4)
v	valve
w	wall
w	water (2.5)
A	area
G	gravitation (3.3)
F	friction

GREEK

α	W/m^2K	convective heat transfer coefficient

α		coefficient belonging to resistance

$$\gamma_A = (\frac{\partial v}{\partial P})_s$$

$$\gamma_B = (\frac{\partial v}{\partial s})_P$$

δ	m	wall thickness
δ		symbol of difference (e.g. $\delta P = P_1 - P_2$)
$\delta(t)$		Dirac (impulse) function
Δ		small difference
∂		partial derivative
ε		emissivity (2.3)
ε	rad/s^2	angular acceleration
ϑ	K	temperature
$\Theta(s)$		Laplace transform of temperature
$\varkappa$		adiabatic exponent
λ		coefficient of flow resistance
λ	W/mK	coefficient of thermal conductivity
λ		eigenvalue
μ		leakage (outflow) coefficient
μ		friction coefficient (2.4)
ν	m^2/s	kinematic viscosity
ξ		damping coefficient
ρ		density
σ		real part of complex variable s
σ	N/m^2	stress (3.3)

$$\tau_P = (\frac{\partial T}{\partial P})_s$$

$$\tau_s = (\frac{\partial T}{\partial s})_P$$

φ	rad	angle
ψ		gas characteristic
ω		imaginary part of complex variable s
ω	rad/s	angular velocity
ω	s^{-1}	natural frequency
$\mathscr{L}, \mathscr{L}^{-1}$		Lapalace transformation and its inverse, respectively

SPECIAL NOTATIONS

DE	differential equation
ODE	ordinary differential equation
PDE	partial differential equation
L	linear
NL	nonlinear
h', h'', r	
ρ', ρ''	saturation variables in Sec. 2.5 and 3.3
v', v'', v'''	

Throughout this whole book we have been consistent in changing, after Laplace transformation, every lower-case letter denoting a time function into an upper-case letter, to represent transformation into the complex s-domain (for example, after transformation $m(t)$ became $M(s)$, $\vartheta(t)$ became $\Theta(s)$ and so on).

A line over a symbol represents the value of the physical variable in the steady state (for example $\bar{P}$, $\bar{v}$ and $\bar{m}$ are pressure, specific volume and mass flow rate in the operating regime in which there is no time change - in the steady operating state). Boldface symbols denote matrices and vectors.

BIBLIOGRAPHY

1. Abgarian, K.A., *Matrichnye i asimptoticheskie metody v teorii lineinyh sistem*, (R), Nauka, Moskva, 1973.

2. Adams, J., Clark, D.R., Louis, J.R., Spanbauer, J.P., *Mathematical Modeling of Once-Through Boiler Dynamics*, IEEE Trans., PAS - 84, pp. 146-156, 1965.

3. Alitshuli, A.D., Kiselev, P.G., *Gidravlika i aerodinamika*, (R), Stroiizdat, Moskva, 1975.

4. Andrews, J.G., McLone, R.R., *Mathematical Modelling*, Butterworths, London, 1976.

5. Aseltine, J.A., *Transform Method in Linear System Analysis*, McGraw-Hill, New York, 1958.

6. Åström, K.J., Bell, R.D., *A Simple Drum Level Model*, Lund Inst. of. Techn., Report LUTFD2/(TFRT-7163)/1-29/ (1979).

7. **Becker**, C., Litz, L., Siffling, G., *Regelungstechnik-Übungsbuch*, (G), AEG-Telefunken, Berlin, Frankfurt a. M., 1982.

8. Bellman, R., *Introduction to Matrix Analysis*, McGraw-Hill, New York, 1970.

9. **Berezovskii**, A.A., *Lekcii po nelineinym kraievym zadacham matematicheskoi fiziki*, chasti I, (R), Naukova dumka, Kiiev, 1976.

10. Blackburn, J.F., Reethof, G., Shearer, J.L., *Fluid Power Control*, The Tech. Press of M.I.T. and John Wiley&Sons, New York, 1960.

11. Bockhardt, H.D., Güntzschel, P., Poetschukat, A., *Grundlagen der Verfahrenstechnik für Ingenieure*, (G), VEB Deutscher Verl. f. Grundstoffindustrie, Leipzig, 1981.

12. Bogomolov, A.I., *Primery gidravlicheskih raschetov*, (R), Transport, Moskva, 1977.

13. Bošnjaković, F., *Nauka o toplini - I dio*, (SC), Tehnička knjiga, Zagreb, 1970.

14. Brack, G., *Dynamische Modelle verfahrenstechnischer Prozesse*, (G), RA 115, VEB Verlag Technik, Berlin, 1972.

15. **Brack**, G., *Dynamik technischer Systeme*, (G), VEB Deutscher Verl. f. Grundstoffindustrie, Leipzig, 1974.

16. **Brack**, G., *Einfache Modelle kontinuirlicher Prozesse*, (G), A. Hüthig Verlag, Heidelberg, 1982.

17. Bronštejn, I.N., Semendjajev, K.A., *Matematički priručnik*, (SC), Tehnička knjiga, Zagreb, 1964.

18. Buckley, P.S., *Techniques of Process Control*, John Wiley&Sons., New York, 1964.

19. **Budak**, B.M., Samarskii, A.A., Tihonov, A.N., *Sbornik zadach po matematicheskoi fizike*, (R), Nauka, Moskva, 1980.

20. Campbell, D.P., *Process Dynamics*, John Wiley, New York, 1958.

21. Chien, K.L., Ergin, E.I., Ling, C., Lee, A., *Dynamic Analysis of a Boiler*, Trans. ASME, Vol. 80, pp. 1809-1819, 1958.

22. **Chaki (Csaki)**, F., *Sovremennaia teoriia upravleniia*, (R), Mir, Moskva, 1975.

23. **Chermak**, I., **Peterka**, V., **Zavorka**, I., *Dinamika reguliruemyh sistem v teploenergetike i himii*, (R), Mir, Moskva, 1972.

24. Debeljković, D.Lj., *Dinamika objekata i procesa*, (SC), Mašinski fakultet, Beograd, 1983.

25. Demidovich, B.P., Maron, I. A., *Computational Mathematics*, Mir, Moskva, 1973.

26. Doležal, R., Varcop, L., *Process Dynamics*, Elsevier, London, 1970.

27. Eckmann, D.P., *Automatic Process Control*, John Wiley&Sons., New York, 1967.

28. Engshuber, M., Müller, R., *Grundlagen der Verfahrenstechnik für Automatisierungsingenieure*, (G), VEB Deutscher Verl. f. Grundstoffindustrie, Leipzig, 1978.

29. Eykhoff, P., *System Identification*, John Wiley&Sons, London, 1974.

30. Fasol, K.H., Jörgl, H.P., *Modelling and Identification*, 5th IFAC-Symp. on Ident. and Par. Estim. - Tutorial, TH Darmstadt, 1979.

31. Findeisen, W., *Grundlagen des Entwurfs von Regelungssystemen*, (G), VEB Verlag Technik, Berlin, 1973.

32. Föllinger, O., *Regelungstechnik*, (G), Elitera, Berlin, 1978.

277

33. Föllinger, O., Franke, D., *Einführung in die Zustandsbeschreibung dynamischer Systeme*, (G), R. Oldenbourg, München, 1982.

34. **Friedly**, J.C., *Dynamic Behavior of Processes*, Prentice-Hall, Englewood Cliffs, N.J., 1972.

35. Fritzsch, W., *Dynamische modelle fertigungstechnischer Prozesse*, (G), RA 169, VEB Verlag Technik, Berlin, 1975.

36. Gilles, E.D., *Systeme mit verteilten Parametern-Einführung in die Regelungstheorie*, (G), Oldenbourg-Verlag, München, 1975.

37. Göldner, K., Kubik, S., *Nichtlineare Systeme der Regelungstechnik*, (G), VEB Verlag Technik, Berlin, 1983.

38. **Guter**, R.S., Janpoliskii, A.R., *Differencialinye uravneniia*, (R), Visshaia shkola, Moskva, 1976.

39. Hanjalić, K., *Dinamika stišljivog fluida*, (SC), Svjetlost, Sarajevo, 1978.

40. Holman, J.P., *Heat Transfer*, McGraw-Hill, New York, 1981.

41. Holmes, R., *The Characteristics of Mechanical Engineering Systems*, Pergamon Press, Oxford, 1977.

42. Isachenko, V.P., Osipova, V.A., Sukomel, A.S., *Teploperedacha*, (R), Energiia, Moskva, 1975.

43. **Isermann**, R., *Theoretische Analyse der Dynamik industrieller Prozesse*, (G), BI-Taschenbuch 764/764 a, Mannheim, 1971.

44. **Isermann**, R., *Anwendung von mathematischen Modellen industrieller Prozesse in der Mess-, Filter-, Regelungs- und Prozessrechentechnik*, (G), VDE- Kongress, München, 1976, VDE-Fachbericht Nr. 29, 1976.

45. Isermann, R., *Dynamik und Regelung industrieller Prozesse*, (G), Lectures-TU Stuttgart, 1977.

46. Kafarov, V., *Cybernetic Methods in Chemistry and Chemical Engineering*, Mir, Moskva, 1976.

47. **Karnop**, D., **Rosenberg**, R., *System Dynamics: A Unified Approach*, John Wiley&Sons, New York, 1975.

48. **Kecman**, V., *Simulation des Regelverhaltens eines aufgeladenen Zwangsdurchlaufdampferzeugers*, (G), IVD Bericht, TU Stuttgart, 1977.

49. **Kecman**, V., *Simulation Analysis of the Dynamics of a Drum Boiler Circulation Loop*, Report No. 80-843-01, Drexel University, Philadelphia, 1980.

50. **Kecman**, V., *Modeliranje i simulaciona analiza dinamičkih procesa u parnom kotlu*, (SC), Ph.D. Thesis, FSB University of Zagreb, 1982.

51. **Kecman**, V., *Modeliranje i analiza dinamičkih svojstava spremnika dvofaznih fluida*, (SC), Proceedings of the JUREMA, Vol. 1, 1983.

52. **Kecman**, V., *Nonlinearity of Dynamic Processes in a Drum Boiler Circulation Loop*, IFAC Workshop, Mod. and Contr. of El. Pow. Pl., Preprints, Como, pp. 47-59, 1983.

53. Kecman, V., *Reducirani model dinamike isparivača*, (SC), Proceedings of the JUREMA, Vol. 1, 1984.

54. **Köhne**, M., *Regelungstheorie örtlich verteilter Systeme*, (G), Lectures-TU Stuttgart, 1976.

55. Kogan, V.B., *Teoreticheskie osnovy tipovyh processov himicheskoi tehnologii*, (R), Himiia, Leningrad, 1977.

56. **Konopacki**, W.A., *The Dynamics and Stability of Evaporation in Steam Generators*, Ph. D. Thesis, Drexel University, Philadelphia, 1979.

57. **Kwatny**, H.G., **McDonald**, J.P., **Spare**, J.H., *A nonlinear model for reheat boiler-turbine-generator systems*, Part II-Development, JACC of the AACC, Paper No. 3-D5, 1971.

58. **Kwatny**, H.G., **Mablekos**, V.E., *The Modeling of Dynamical Processes*, IEEE Conf., Paper No. TA3-1, 1975.

59. Leleev, N.S., *Neustanovivsheesia dvizhenie teplonositelia v obogrevaemyh trubah moschnyh parogeneratorov*, (R), Energiia, Moskva, 1978.

60. McClamroch, N.H., *State Models of Dynamic Systems*, Springer-Verlag, New York, 1980.

61. **Merkin**, D.R., *Vvedenie v teoriiu ustoichivosti dvizheniia*, (R), Nauka, Moskva, 1976.

62. Nicholson, H., *Modelling of Dynamical Systems*, Vol. 1, Peregrinus Ltd., London, 1980.

63. Novaković B., *Opći model strukturne sinteze složenih dinamičkih sistema*, (SC), Proceedings of the JUREMA, Vol. 2, pp. 131-135, 1978.

64. Ogata, K., *State Space Analysis of Control Systems*, Prentice-Hall, Englewood Cliffs, N.J., 1967.

65. **Ogata** K., *System Dynamics*, Prentice-Hall, Englewood Cliffs, N.J., 1978.

66. Peterka, V., *Analytische Ermittlung der Dampfdruckdynamik in Zwangsdurchlaufkesseln*, (G), MSR 7, pp. 229-239, 1964.

67. Petrovački, D., *Upravljanje sistemima sa raspodijeljenim parametrima*, (SC), Proceedings of the JUREMA, Vol. 1, pp. 5-8, 1984.

68. **Pontriagin**, L.S., *Obyknovennye differencialinye uravneniia*, (R), Nauka, Moskva, 1974.

69. **Profos**, P., *Die Regelung von Dampfanlagen*, (G), Springer-Verlag, Berlin, 1962.

70. Profos, P., *Modellbildung und ihre Bedeutung in der Regelungstechnik*, (G), VDI-Berichte Nr. 276, 1977.

71. Ray, A., Bowman, H.F., *A Nonlinear Dynamic Model of a Once-through Subcritical Steam Generator*, J. of Dyn. Syst., Meas. and Contr., ASME Trans., pp. 332-339, 1976.

72. Ray, W.H. *Advanced Process Control*, McGraw-Hill, New York, 1981.

73. Reinisch, K., *Kybernetische Grundlagen und Beschreibung kontinuirlicher Systeme*, (G), VEB Verlagen Technik, Berlin, 1974.

74. **Reynolds**, W.C., *Thermodynamics*, McGraw-Hill, New York, 1968.

75. **Serov**, E.P., **Korolikov**, B.P., *Dinamika parogeneratorov*, (R), Energiia, Moskva, 1972.

76. Shapiro, A.H., *The Dynamics and Thermodynamics of Compressible Fluid Flow*, The Ronald Press Company, New York, 1953.

77. Streic, V., *Stavová teorie lineárniho diskrétniho řizeni*, (CS), Academia, Praha, 1978.

78. Shumskaia, L.S., *The Variation of the Major Parameters in Drum Boilers in Transient Conditions*, in Problem of Heat Transfer and Hydraulics of Two-Phase Media, Kutateladze, S.S., Pergamon Press, 1969.

79. Šurina, T., *Automatska regulacija*, (SC), Školska knjiga, Zagreb, 1981.

80. **Takahashi**, Y., **Rabins**, M.J., **Auslander**, D.M., *Control and Dynamic Systems*, Addison-Wesley, Reading, Mass., 1972.

81. Teichmann, W., *Angewandte Anlagenautomatisierung*, (G), VEB Verlag Technik, Berlin, 1983.

82. **Tihonov**, A.N., **Samarskii**, A.A., *Uravneniia matematicheskoi fiziki*, (R), Nauka, Moskva, 1977.

83. Töpfer, H., Rudert, S., *Einführung in die Automatisierungstechnik*, (G), VEB Verlag Technik, Berlin, 1976.

84. **Varcop**, L., *Dynamika odpařovaciho pásma výparniku s jedorozměrnym průtokem*, (CS), Rozpravy CSAV, Praha, 1967.

85. VDI/VDE-GMR, (Editor Isermann, R.), *Prozessmodell-Katalog*, (G), VDI-Verlag, Düsseldorf, 1976.

86. **Voronov**, A.A., *Ustoichivost, upravliaemost, nabliudaemost*, (R), Nauka, Moskva, 1979.

87. **Weber**, T.W., *An Introduction to Process Dynamics and Control*, John Wiley&Sons, New York, 1973.

The names of the authors whose books are the closest in content or in approach to specific parts of this book have been printed in boldface.

(CS ... in Czech, G ... in German, R ... in Russian, SC ... in SerboCroatian)

Lecture Notes in Control and Information Sciences

Edited by M. Thoma and A. Wyner

Lecture Notes in Control and Information Sciences

Edited by M. Thoma and A. Wyner

Lecture Notes in Control and Information Sciences

Edited by M. Thoma and A. Wyner